Lecture Notes
in Business Information Processing   297

Series Editors

Wil M.P. van der Aalst
  Eindhoven Technical University, Eindhoven, The Netherlands
John Mylopoulos
  University of Trento, Trento, Italy
Michael Rosemann
  Queensland University of Technology, Brisbane, QLD, Australia
Michael J. Shaw
  University of Illinois, Urbana-Champaign, IL, USA
Clemens Szyperski
  Microsoft Research, Redmond, WA, USA

More information about this series at http://www.springer.com/series/7911

Josep Carmona · Gregor Engels
Akhil Kumar (Eds.)

Business Process Management Forum

BPM Forum 2017
Barcelona, Spain, September 10–15, 2017
Proceedings

 Springer

Editors
Josep Carmona
Department of Computer Science
Universitat Politècnica de Catalunya
Barcelona
Spain

Gregor Engels
Department of Computer Science
Paderborn University
Paderborn
Germany

Akhil Kumar🆔
Department of Supply Chain and
   Information Systems
Pennsylvania State University
University Park, PA
USA

ISSN 1865-1348   ISSN 1865-1356 (electronic)
Lecture Notes in Business Information Processing
ISBN 978-3-319-65014-2   ISBN 978-3-319-65015-9 (eBook)
DOI 10.1007/978-3-319-65015-9

Library of Congress Control Number: 2017948177

Printed on acid-free paper

This Springer imprint is published by Springer Nature
The registered company is Springer International Publishing AG
The registered company address is: Gewerbestrasse 11, 6330 Cham, Switzerland

Preface

The International Conference on Business Process Management (BPM), now in its 15th year, has clearly established itself as the most important academic event in BPM. It is the premium event for researchers, practitioners, and developers in this area. In 2016, the Steering Committee of this conference created a new sub-track, called the BPM Forum, to be held in conjunction with the main conference and with a separate set of proceedings. It made its successful debut at the Rio conference, and, hence, it was also included in the conference held in Barcelona during September 10–15, 2017.

The aim of the BPM Forum is to host innovative research which has high potential of stimulating discussion, but does not quite meet the rigorous quality criteria for the main research track. The papers selected for the forum are expected to showcase fresh ideas from exciting and emerging topics in BPM, even if they were not yet at the same level of maturity as the regular papers at the conference. We picked these papers from those that could not be accepted to BPM 2017 based on the recommendation of the Program Committee (PC) members who were assigned to evaluate them. As far as possible, we avoided overlap with the workshops associated with the BPM Conference.

Consequently, we selected 11 high-quality papers out of 97 papers that were not included in the main program of BPM. The papers in this volume cover topics related to process models and metrics, mining and compliance, and other innovative ideas such as gamification, smart devices, and digital innovation as they pertain to BPM. The review process involved 21 senior PC members and 103 regular PC members. Each paper was reviewed by a team comprising a senior PC and four regular PC members who engage in a discussion phase after the initial reviews are prepared. The authors receive four review reports, and a meta-review that summarizes the reviews and the discussion.

We are grateful for the generous support of the sponsors of the BPM conference: Signavio, Celonis, IBM, Diputacio de Tarragona, MyInvenio, DCR, Bizagi, CA Technologies, Mysphera, and Springer. We very much hope you enjoy reading the papers in this volume.

September 2017

Josep Carmona
Gregor Engels
Akhil Kumar

Organization

BPM 2017 Forum was a sub-track of BPM 2017, which was organized by the Universitat Politècnica de Catalunya, and took place in Barcelona, Spain.

Steering Committee

Wil van der Aalst (Chair)	Eindhoven University of Technology, The Netherlands
Boualem Benatallah	University of New South Wales, Australia
Jörg Desel	University of Hagen, Germany
Schahram Dustdar	Vienna University of Technology, Austria
Marlon Dumas	University of Tartu, Estonia
Manfred Reichert	University of Ulm, Germany
Stefanie Rinderle-Ma	University of Vienna, Austria
Barbara Weber	Technical University of Denmark, Denmark
Mathias Weske	HPI, University of Potsdam, Germany
Michael zur Muehlen	Stevens Institute of Technology, USA

Executive Committee

Conference Chair

Josep Carmona	Universitat Politècnica de Catalunya, Spain

Program Chairs

Josep Carmona	Universitat Politècnica de Catalunya, Spain
Gregor Engels	Paderborn University, Germany
Akhil Kumar	Penn State University, USA

Industry Chairs

Marco Brambilla	Politecnico Milano, Italy
Thomas Hildebrandt	IT University of Copenhagen, Denmark
Victor Muntès	CA Technologies, Spain
Darius Silingas	No Magic Europe and ISM UME, Lithuania

Workshops

Matthias Weidlich	Humboldt-Universität zu Berlin, Germany
Ernest Teniente	Universitat Politècnica de Catalunya, Spain

Tutorial and Panel Chairs

Joaquin Ezpeleta	University of Zaragoza, Spain
Dirk Fahland	Eindhoven University of Technology, The Netherlands
Barbara Weber	Technical University of Denmark, Denmark

Demo Chairs

Robert Clarisó	Universitat Oberta de Catalunya, Spain
Henrik Leopold	VU University Amsterdam, The Netherlands

Doctoral Consortium Chairs

Antonio Ruiz Cortés	University of Seville, Spain
Mathias Weske	HPI, University of Potsdam, Germany

Publicity Chairs

Jordi Cabot	Open University of Catalonia, Spain
Marcos Sepúlveda	Pontificia Universidad Católica de Chile, Chile
Marco Montali	Free University of Bozen-Bolzano, Italy

Sponsorship Chairs

Carlos Fernandez-Llatas	Universidad Politecnica de Valencia, Spain
Pedro Álvarez	University of Zaragoza, Zaragoza
Rubén Mondéjar	Universitat Rovira i Virgili, Spain

Co-located Events Chairs

Manuel Lama	University of Santiago de Compostela, Spain
Alberto Manuel	Microsoft, Lisbon
Antonio Valle	G2, Spain

Web and Social Media Chairs

Jorge Munoz-Gama	Pontificia Universidad Católica de Chile, Chile
Andrea Burattin	University of Innsbruck, Austria

Proceedings Chair

Alexander Teetz	Paderborn University, Germany

Senior Program Committee

Marlon Dumas	University of Tartu, Estonia
Schahram Dustdar	TU Wien, Austria
Avigdor Gal	Technion, Israel
Richard Hull	IBM T.J. Watson Research Center, USA
Fabrizio Maria Maggi	University of Tartu, Estonia
Massimo Mecella	Sapienza Università di Roma, Italy

Jan Mendling	Wirtschaftsuniversität Wien, Austria
Marco Montali	Free University of Bozen-Bolzano, Italy
Artem Polyvyanyy	Queensland University of Technology, Australia
Manfred Reichert	University of Ulm, Germany
Hajo A. Reijers	Vrije Universiteit Amsterdam, The Netherlands
Stefanie Rinderle-Ma	University of Vienna, Austria
Michael Rosemann	Queensland University of Technology, Australia
Antonio Ruiz-Cortés	University of Seville, Spain
Pnina Soffer	University of Haifa, Israel
Jianwen Su	University of California at Santa Barbara, USA
Boudewijn Van Dongen	Eindhoven University of Technology, The Netherlands
Irene Vanderfeesten	Eindhoven University of Technology, The Netherlands
Barbara Weber	Technical University of Denmark, Denmark
Matthias Weidlich	Humboldt-Universität zu Berlin, Germany
Mathias Weske	HPI, University of Potsdam, Germany

Program Committee

Mari Abe	IBM Research, Japan
Shivali Agarwal	IBM, India Research Lab, India
Ahmed Awad	Cairo University, Egypt
Hyerim Bae	Pusan National University, South Korea
Bart Baesens	KU Leuven, Belgium
Seyed-Mehdi-Reza Beheshti	University of New South Wales, Australia
Boualem Benatallah	University of New South Wales, Australia
Giorgio Bruno	Politecnico di Torino, Italy
Joos Buijs	Eindhoven University of Technology, The Netherlands
Andrea Burattin	University of Innsbruck, Austria
Jorge Cardoso	University of Coimbra, Portugal
Fabio Casati	University of Trento, Italy
Jan Claes	Ghent University, Belgium
Florian Daniel	Politecnico di Milano, Italy
Massimiliano de Leoni	Eindhoven University of Technology, The Netherlands
Jochen De Weerdt	KU Leuven, Belgium
Patrick Delfmann	European Research Center for Information Systems (ERCIS), Germany
Jörg Desel	University of Hagen, Germany
Alin Deutsch	University of California San Diego, USA
Chiara Di Francescomarino	Fondazione Bruno Kessler-IRST, Italy
Remco Dijkman	Eindhoven University of Technology, The Netherlands
Dirk Draheim	Tallinn University of Technology, Estonia
Johann Eder	Alpen Adria Universität Klagenfurt, Austria
Rik Eshuis	Eindhoven University of Technology, The Netherlands
Joerg Evermann	Memorial University of Newfoundland, Canada

Dirk Fahland	Technische Universiteit Eindhoven, The Netherlands
Marcelo Fantinato	University of São Paulo, Brazil
Peter Fettke	DFKI/Saarland University, Germany
Hans-Georg Fill	University of Vienna, Austria
Walid Gaaloul	Télécom SudParis, France
Luciano García-Bañuelos	University of Tartu, Estonia
Christian Gerth	Osnabrück University of Applied Sciences, Germany
Chiara Ghidini	FBK-irst, Italy
María Teresa Gómez-López	University of Seville, Spain
Guido Governatori	Data61, Australia
Sven Graupner	Hewlett-Packard Laboratories, USA
Paul Grefen	Eindhoven University of Technology, The Netherlands
Daniela Grigori	University of Paris-Dauphine, France
Thomas Hildebrandt	IT University of Copenhagen, Denmark
Mieke Jans	Hasselt University, Belgium
Anup Kalia	IBM T.J. Watson Research Center, USA
Dimka Karastoyanova	Kühne Logistics University, Germany
Ekkart Kindler	Technical University of Denmark, Denmark
Agnes Koschmider	Karlsruhe Institute of Technology, Germany
John Krogstie	Norwegian University of Science and Technology, Norway
Jochen Kuester	Bielefeld University of Applied Sciences, Bielefeld
Marcello La Rosa	Queensland University of Technology, Australia
Geetika Lakshmanan	IBM T.J. Watson Research Center, USA
Manuel Lama Penin	University of Santiago de Compostela, Spain
Alexei Lapouchnian	University of Toronto, Canada
Ralf Laue	University of Applied Sciences Zwickau, Germany
Henrik Leopold	VU University Amsterdam, The Netherlands
Rong Liu	IBM Research, USA
Irina Lomazova	National Research University Higher School of Economics, Russia
Peter Loos	DFKI/Saarland University, Germany
Heiko Ludwig	IBM Research, USA
Hamid Motahari	IBM Research, USA
Juergen Muench	Reutlingen University, Germany
John Mylopoulos	University of Toronto, Canada
Nanjangud Narendra	Ericsson Research Bangalore, India
Selmin Nurcan	Université Paris 1 Panthéon-Sorbonne, France
Hye-Young Paik	University of New South Wales, Australia
Oscar Pastor Lopez	Universitat Politecnica de Valencia, Spain
Dietmar Pfahl	University of Tartu, Estonia
Geert Poels	Ghent University, Belgium
Frank Puhlmann	Bosch Software Innovations, Germany
Mu Qiao	IBM Almaden Research Center, USA

Jan Recker	Queensland University of Technology, Australia
Manuel Resinas	University of Seville, Spain
Maximilian Roeglinger	FIM Research Center, Germany
Shazia Sadiq	The University of Queensland, Australia
Flavia Santoro	Federal University of the State of Rio de Janeiro, Brazil
Rainer Schmidt	Munich University of Applied Sciences, Germany
Heiko Schuldt	University of Basel, Switzerland
Marcos Sepúlveda	Pontificia Universidad Católica de Chile, Chile
Quan Z. Sheng	Macquarie University, Australia
Renuka Sindhgatta	IBM Research, India
Sergey Smirnov	SAP Research, Germany
Marc Sole	CA Strategic Research Labs, CA Technologies, Spain
Minseok Song	Pohang University of Science and Technology, South Korea
Harald Störrle	Danmarks Tekniske Universitet, Denmark
Heiner Stuckenschmidt	University of Mannheim, Germany
Keith Swenson	Fujitsu, USA
Samir Tata	IBM Research, USA
Pankaj Telang	SAS Institute Inc., USA
Ernest Teniente	Universitat Politècnica de Catalunya, Spain
Arthur Ter Hofstede	Queensland University of Technology, Australia
Lucinéia Heloisa Thom	Federal University of Rio Grande do Sul, Brazil
Farouk Toumani	LIMOS/Blaise Pascal University, France
Peter Trkman	University of Ljubljana, Slovenia
Roman Vaculín	IBM T.J. Watson Research Center, USA
Wil van der Aalst	Eindhoven University of Technology, The Netherlands
Amy Van Looy	Ghent University, Belgium
Jan Vanthienen	KU Leuven, Belgium
Hagen Voelzer	IBM Research, Zurich, Switzerland
Jianmin Wang	Tsinghua University, China
Ingo Weber	Data61, CSIRO, Australia
Lijie Wen	Tsinghua University, China
Karsten Wolf	Universität Rostock, Germany
Moe Wynn	Queensland University of Technology, Australia
Liang Zhang	Fudan University, China

Additional Reviewers

Alexander Norta	Javier de San Pedro	Riccardo De Masellis
Bernardo Nugroho Yahya	Johannes De Smedt	Rick Gilsing
Carlos Rodriguez	Julius Köpke	Seyed-Mehdi-Reza
Chun Ouyang	Marigianna Skouradaki	Beheshti
David Sanchez-Charles	Mauro Dragoni	Toon Jouck
Erik Proper	Mirela Madalina Botezatu	Sander Peters
Gert Janssenswillen	Montserrat Estañol	Wasana Bandara
Jaehun Park	Pavlos Delias	

Sponsors

Platinum Sponsor

Gold Sponsor

Gold Sponsor

Silver Sponsor

Bronze Sponsor

Bronze Sponsor

Bronze Sponsor

Demo Sponsor

Contents

Models and Metrics

Elements for Tailoring a BPM Maturity Model to Simplify its Use   3
    Marie-Therese Christiansson and Amy Van Looy

A New Framework for Defining Realistic SLAs:
An Evidence-Based Approach   19
    Minsu Cho, Minseok Song, Carlos Müller, Pablo Fernandez,
    Adela del-Río-Ortega, Manuel Resinas, and Antonio Ruiz-Cortés

A Template for Categorizing Business Processes in Empirical Research.   36
    Daniel Lübke, Ana Ivanchikj, and Cesare Pautasso

Mining and Compliance

Toward a New Generation of Log Pre-processing Methods
for Process Mining   55
    Paolo Ceravolo, Ernesto Damiani, Mohammadsadegh Torabi,
    and Sylvio Barbon Jr.

A Taxonomy of Compliance Processes for Business Process Compliance. . . .   71
    Tobias Seyffarth, Stephan Kühnel, and Stefan Sackmann

Improving Pattern Detection in Healthcare Process Mining
Using an Interval-Based Event Selection Method   88
    Amirah Alharbi, Andy Bulpitt, and Owen Johnson

Soundness of Decision-Aware Business Processes.   106
    Kimon Batoulis and Mathias Weske

BPM Miscellany

BPMS-Game: Tool for Business Process Gamification.   127
    Javier Mancebo, Felix Garcia, Oscar Pedreira,
    and Maria Angeles Moraga

Events in Business Process Implementation: Early Subscription
and Event Buffering   141
    Sankalita Mandal, Matthias Weidlich, and Mathias Weske

Artifact-Driven Monitoring for Human-Centric Business Processes
with Smart Devices: Assessment and Improvement   160
  Giovanni Meroni and Pierluigi Plebani

A Quantitative Study of the Link Between Business Process
Management and Digital Innovation   177
  Amy Van Looy

Author Index   193

Models and Metrics

Models and Methods

Elements for Tailoring a BPM Maturity Model to Simplify its Use

Marie-Therese Christiansson[1](✉) and Amy Van Looy[2]

[1] Department of Information Systems,
Karlstad Business School, Karlstad University,
Universitetsgatan 2, 651 88 Karlstad, Sweden
marie-therese.christiansson@kau.se
[2] Department of Business Informatics and Operations Management,
Faculty of Economics and Business Administration, Ghent University,
Tweekerkenstraat 2, 9000 Ghent, Belgium
Amy.VanLooy@UGent.be

Abstract. Although research exists on Business Process Management (BPM) maturity models, few studies report on their practical use. This paper explores the situational needs and practitioner's views on assessing BPM maturity. Data triangulation uncovered different applications in Swedish industry and public sector organizations through three phases: (1) data collection in a practitioner-driven BPM maturity model design, (2) validation of the design in a workshop and follow-up interviews, and (3) testing the BPM maturity model by practitioners. The basic assumption is that a generic BPM maturity model will most likely not fit all organizations. Therefore, a framework is presented with elements (e.g. scope and measures) for a BPM maturity model to be customized to an organization's needs, supplemented by a practical 'tailoring template'. The framework and template contribute to the BPM discipline with a Swedish example, and allows twelve design propositions with recommendations to simplify the application of BPM maturity models and enhance their fit.

Keywords: Business Process Management · Lifecycle management · Maturity model · Adoption and practice · Tailoring · Customization

1 Introduction

Maturity models (MMs) are considered important in the Business Process Management (BPM) discipline, and more specifically in process lifecycle management. The benefits of using BPM are, among others, increased customer satisfaction, greater efficiency and cost savings, more transparency of activities, business agility and compliance ease [27, 32]. Awareness is growing that BPM requires an integrated and holistic lifecycle approach [30], also called Business Process Orientation (BPO). The latter indicates that, in addition to methods and IT, core capabilities in terms of strategic alignment, governance, people, and culture [21, 33] are increasingly highlighted in the process lifecycle of identifying, designing, analyzing, implementing, measuring and improving the business processes in organizations [19]. Thus, BPM covers aspects ranging from process characteristics to the

© Springer International Publishing AG 2017
J. Carmona et al. (Eds.): BPM Forum 2017, LNBIP 297, pp. 3–18, 2017.
DOI: 10.1007/978-3-319-65015-9_1

organizational structure required for process-oriented work as well as people working with a customer-focused mind-set [28]. A BPM maturity model can help organizations achieve the promised benefits by gradually addressing all relevant BPO characteristics. We follow Becker et al. [1] to define a BPM maturity model as a sequence of discrete maturity levels or steps with representations in an anticipated, desired, or evolutionary path for business processes, and at the same time as a plan for work practices and intended effects of BPO. Maturity, in differently formulated definitions, is used as a measure to evaluate an organization in terms of how advanced the organization is in the area of BPO or what has been achieved in this area [30]. One of the best-known maturity models is CMMI, but many others exist [11, 24]. However, while BPM maturity model designs have gained considerable attention in research [18, 19], research on the practitioner's ability to actually use maturity models has been limited and even neglected so far [29].

In fact, due to a large number of BPM maturity models and despite the practical uptake and potential benefits of maturity models like CMMI [24], some practitioners still seem reluctant to use BPM maturity models [36]. Since there are hundreds of maturity models [24], the first barrier for practitioners is to find the appropriate BPM maturity models to select from. Van Looy et al. [29] identified 14 criteria that potential users should consider when selecting a BPM maturity model. Additionally, the present study intends to explore different applications or practitioners' uses once a particular maturity model has been chosen. As such, our findings will be related to prior research on what to consider when selecting a BPM maturity model [22, 27, 29], and on further defining what makes a BPM maturity model useful [18]. More specifically, researchers and practitioners have proposed maturity models with a varied focus, scope and depth [22, 27, 35], as well as with a diverse construction and different target groups [9, 19], which makes it difficult to identify the most appropriate model depending on situational needs (e.g. regarding a specific business context, organizational characteristics and objectives). Furthermore, domain-specific BPM maturity models are proposed which are tailored to a specific application area (e.g. a public administration) and a specific purpose (e.g. the fulfillment of a 48-h-service promise) [8]. Tarhan et al. [25] stress that methodical applications of many generic BPM MMs are lacking, e.g. design principles for a prescriptive use are largely unmet as well as empirical studies to demonstrate the validity and usefulness of BPM maturity models. Only a handful of studies examine the adoption of these maturity models and their effects on business performance [25, 26].

Moreover, vom Brocke et al. [31] claim that BPM needs to consider the business context in which it is applied, i.e. with factors related to BPM goals, processes, the organization, and the environment. Some of these factors should be considered in both a process improvement project and continuous business development, such as process types (management/core/support), process characteristics, corporate culture, resources, and the competiveness and uncertainty in the environment to affect the management practices. Context factors may also change over time and influence the BPM requirements as well as the maturity criteria in a given situation. However, maturity models are generally criticized for their one-size-fits-all approach [31] i.e. proposing a predefined "end state" without empirical foundation in reality [22], and missing assistance to help users determine the current stage of maturity and realize progression to the next stage [5]. While the context factors by vom Brocke et al. [31] are related to

BPM goals, processes, the organization, and the environment, better support is needed when applying them to the actual improvement of work practices.

To fill this gap, our study elaborates on the possibilities of tailoring a generic BPM maturity model, once it has been chosen, namely to make modifications taking into account the situational needs in a specific context (i.e. the use case or setting), as well as the organizational pre-conditions for conducting an assessment. Tailoring might range from minor 'tweaking' to major customization of a maturity model for a specific target group [12]. For instance, decisions on the next step/stage may depend on an analysis of motivation (e.g. alignment to goals or a cost-benefit analysis) taking into account the business context, where the competition is heading, or what customers expect. Our study is empirically driven with an inductive approach in close cooperation with practitioners to explore such applications or situational needs in organizations [17] and identify key elements in tailoring BPM maturity models.

In the remainder of this article, we first identify challenges for using a BPM maturity model based on the literature (Sect. 2). The empirical research design with multiple methods for data collection is presented in Sect. 3. Then, lessons learned from practitioners' applications are presented (Sect. 4). Section 5 discusses a 'tailoring template' for BPM maturity models based on key elements and design propositions, while Sect. 6 concludes with limitations and future research directions.

2 Theoretical Background

Maturity is not a straightforward 'journey', but rather steps up and/or down the stairs [4, 30] in a so-called 'process staircase' that may illustrate the degree of maturity. For instance, possible steps are as follows:

- **Step 1:** processes are mapped – BPM starts with identifying and describing processes and the organization becomes aware of its business process performance.
- **Step 2:** processes are established – BPM is established with roles and responsibilities and makes sure that process descriptions are easy to store, find and access. The organization now has a process description guide with communication and decisions across administrations and/or organizations.
- **Step 3:** processes are evaluated and improved – BPM teams evaluate the value contribution in processes based on metrics and quality indicators to clarify what is needed to achieve the business goals or objectives in the organization. Continuous process improvements may follow performance and quality demands.
- **Step 4:** processes are implemented and the organization is structured and managed by processes. Employees and systems work horizontally towards an external customer, and employees know their role and can act better. Hence, BPO is achieved.

In the literature, we identified some characteristics that vary for (BPM) maturity models. Such variations may be experienced as challenging for practitioners when using a BPM maturity model, i.e. in their role of maturity model users (e.g. business developers or process managers) and respondents (e.g. regular employees and managers). We identified eight challenges (C) for practitioners in the literature:

- **C1:** To find MMs using appropriate channels
- **C2:** To know the intended target group and related value propositions in use
- **C3:** To choose a particular MM that best fits a purpose
- **C4:** To know the scope of a particular MM
- **C5:** To know the degree to which the BPO foundation is implemented
- **C6:** To understand what to measure in each step or stage of a particular MM
- **C7:** To understand how the outcomes of maturity are measured in a MM
- **C8:** To understand how to decide on appropriate respondents for a MM

The first challenge for practitioners is to find maturity models (C1) in a particular domain or field (e.g. business excellence in general or related to supply chain management, public administrations or software development), and to know through which channels a relevant maturity model can be found (e.g. in a book, an article, in conference proceedings, or on a website). Practitioners may experience difficulties to know where to look and/or how to understand these findings. For instance, a maturity model might be described in different versions and in different publications, some open to access while others are restricted to an association membership. Thus, it is difficult to reach an entire maturity model description, and nearly impossible to know when a description is complete. One way to address this challenge is by offering a selection tool [29], which might be broadened towards tailoring once a maturity model has been chosen. Practitioners should also be aware that such a selection tool exists. Nonetheless, selection criteria differ from tailoring criteria. For instance, the 14 selection criteria (SC) of Van Looy et al. [29] are limited to model-related characteristics, as defined by a maturity model itself, and form the basis on which tailoring can start. For instance, the user's ability to find the right *channel* for BPM maturity models (C1) and select an appropriate maturity model is related to all selection criteria of Van Looy et al. In addition, some other selection criteria are also related to challenges due to the target organization per se, i.e. the business process types (SC4), data collection techniques (SC6), number of business processes (SC9), assessment duration (SC10), number of assessment items (SC13) and direct costs (SC14).

Secondly, the challenge to identify the target group (e.g. a business developer, process manager or CEO) for using a particular maturity model and its value propositions (C2) is partly related to the assessment availability (SC11) and validation (SC8) of a maturity model. Only few maturity models have explicit information available for practitioners (e.g. by offering a survey tool, c.f. [9]), or allow customizing based on user profiles. In addition, different users will experience different values of using a maturity model. Hence, developing and offering a BPM maturity model can be described as the creation of value propositions (i.e. possibilities and benefits) by the configurations of several different practices and resources that create value, e.g. service innovation [23]. Nonetheless, value created by using a maturity model must be evaluated from the perspective of user experience in knowing if the maturity model also generates attractive value propositions with tangible and intangible benefits [23], besides the intended effects with BPO. Even if research on process-oriented organizations is still relatively recent, research on the effects of BPO [15] and BPM efforts on improved business performance are reported [15, 16], as well as on mature business processes [26]. The latter studies were conducted in organizations using BPM maturity models to state, in our words, the maturity model's value propositions. As such, researchers should confirm a relation between using a maturity model and better results, which is highly relevant since the target group needs to understand the rationale and effects on business performance when using a particular maturity model.

Thirdly, the choice of a particular maturity model that best fits a *purpose* (C3 and SC7) builds upon the notion of 'capability maturity' in line with the CMMI tradition for software processes, and the practitioners' purposes, namely to describe (i.e. As-Is, enterprise readiness), to prescribe (i.e. To-Be, road map where and how to improve), and/or to compare (i.e. benchmarking) [18, 22].

The ultimate aim of using a BPM maturity model is to measure progress towards the intended effects in BPO. In previous research, different notions of *scope* (C4) in relation to progression are found, namely business process maturity [27, 29], business process modeling maturity [9], business process management maturity [19], as well as the holistic and organization-wide definition to increase business process orientation maturity [27]. In each scope, a number of outlined capabilities are based on pre-defined levels, steps, stages or degrees to reach a certain state of maturity [18, 22]. Thus, a challenge for practitioners is to know the scope of a particular maturity model at hand [22], which relates to the selection criteria for capabilities (SC1) and architecture type (SC2). In particular, when climbing the 'process staircase', a different scope might be in mind when respondents answer the assessment questions or statements. In each step or stage, maturity may concern the operational business processes, the BPM work practice per se or how mature employees are in terms of a BPO mind-set and actions.

Following the 'process staircase', another challenge for practitioners is to know which *foundations of BPO* (e.g. an end-to-end focus, customer-driven improvements, etc.) are actually covered (C5). The fifth challenge thus refers to capabilities (SC1) or areas of interest. For instance, Van Looy et al. [27, 28] present a capability framework with (sub) capabilities based on the traditional process lifecycle (i.e. process modeling, deployment, optimization and management), supplemented by a process-oriented culture and structure. Rosemann and de Bruin [20] provide an alternative capability framework with (sub) components, i.e. IT, strategic alignment, governance, methods, people, and culture [6, 7, 21]. Variants on capabilities or areas of interest are motivation, governance, modeling, tool administration, library management, stakeholder management and training [9]. Similarly, Rohloff [19] introduces several categories that have an impact on BPM success, while Indulska et al. [14] describe different challenges and issues within process modeling. Further on, research on principles of good BPM [32] and critical success factors for adopting BPM [2] add important content to be measured in a BPM maturity model. Hence, it is difficult for practitioners to know if and which maturity models are implementing *research results on challenges and success factors* to improve BPM work practice.

The sixth challenge emphasizes that the outlined capabilities are dependent on and thus measured according to pre-defined steps or stages [18, 22], representing the degree of BPO [4, 30], namely ranging from basic to more advanced capabilities. For practitioners, the challenge is to understand (cf. SC1: capabilities and SC3: architecture details) *what to measure in each step or stage* (C6) of a particular maturity model. In addition, the content in each step or stage may refer to tasks in the BPM work practice (i.e. concrete steps to take) or intended effects (i.e. stage to achieve) in BPO. Thus, *multiple dimensions* occur according to the degree of BPO, as well as in scope and BPO foundations as discussed above, with BPM maturity models referring to terms such as 'levels', 'range', 'meaning', 'state' and 'performance'.

Until now, the previous challenges illustrate that BPM is a complex domain with different aspects (such as capabilities, components, areas, issues, categories for success, challenges). It thus seems hard for a practitioner to grasp and manage the BPM environment, namely what matters (challenges, success factors and issues), what to provide (capabilities), what to do (actions to take in process work) and what to achieve (visions and goals in an organization as well as intended effects with BPO). The subsequent challenges are related to what is measured as outcomes of maturity, and by whom maturity is measured, i.e. the selection of respondents.

The seventh challenge considers *how to measure the outcomes* (C7) of maturity [32], i.e. to quantify progress and performance (cf. SC5: rating scale and SC8: validation). Choong [3] has identified major weaknesses in previous research concerning BPM measurement (e.g. not properly matched with product or service quality, nor a stakeholder perspective). Performance information is still largely financial (e.g. accounting), while many types of performance should be measured qualitatively (e.g. customer satisfaction). Additionally, Choong [3] stresses the fact that the question remains elusive as to whom, and for what purpose the measured information ought to be communicated. Hence, as BPM is critical for organization management, metrics need to be identified according to the purpose and intentions of BPO, the desired effects and value of process improvements and process work. In particular, maturity models typically measure maturity based on how organizations proceed, i.e. their degree of performance when developing BPM capabilities [3, 20] or to what degree the effects in BPO are intended and implemented [11]. Accordingly, a particular maturity model should include recommendations in the sense of explications and advice focusing on criteria, objectives, variables, and measures [18] to know how to move up the stairs.

The final challenge involves the selection of *appropriate respondents* (C8) who should have a certain degree of insight into BPM work practice and knowledge of BPO. If practitioners are not familiar with the BPO foundations [29, 32] and the ongoing process practice, it would seem complicated to assess the BPM maturity in an organization. This challenge builds upon zur Muehlen's report on 'a little knowledge is a dangerous thing' [37].

In sum, while prior work on selection criteria and maturity model design indicates possible variations among BPM maturity models, more research is needed on specific *tailoring criteria* to assist practitioners using those models. For instance, tailoring criteria may supplement prior efforts by specifically being assessment-related in terms of pre-conditions in the organization, planning for the assessment context, knowledge and ability by users and respondents.

3   Research Design

Research on BPM maturity models is mainly tested within organizations and accordingly modified based on practice and academic responses [19, 29]. Limited studies observe how practitioners experience the use of a BPM maturity model in their local practice [38], apart from participating in a research project with guidance from scholars or consultants. For instance, practitioners may lack a researcher's understanding of the BPO foundation and generic BPM maturity models, their intended use, and the meaning of central concepts. To learn from practice and the needs for support when using a maturity model, the current study follows a Practice Research approach [10], namely a situational inquiry (what practitioners say they do and what they actually do) based on the knowledge domain of using BPM maturity models as a common interest shared by researchers and practitioners.

The data collection is based on an in-depth study of work practices, including observations, interventions and reflective learning. The analysis is performed based on the design-science paradigm [13] with general design phases to create a framework that helps identify the key elements for tailoring a BPM maturity model. The study relies on close collaboration with Swedish practitioners in local BPM practices to derive lessons learned for tailoring BPM maturity models, which can help theorizing on generic design artifacts within the BPO domain. A series of qualitative and interpretive studies [34] were conducted for data triangulation in three phases:

- **Data collection** in a practitioner-driven BPM maturity model design (8 practitioners in a pre-study and 36 course participants),
- **Validation** of a BPM maturity model design in a workshop (7 practitioners) and follow-up interviews (5 practitioners), and
- **Testing** of the BPM maturity model (17 practitioners).

The selection of practitioners was not done by the researchers, but based on a broader inquiry. Organizations were invited on the basis of their relation to the university by the first author during a five-year time frame in terms of participation in courses, networks and workshops. The final selection of practitioners was done by the organizations themselves. In total, *73 practitioners* from *23 organizations* (e.g. municipalities, universities, a bank, the tax authority, a management consultancy firm and IT consultancy firms, an insurance company, and the Swedish tenant association) participated in the joint tailoring of a practitioner-driven BPM maturity model, called the PoP (Process Orientation in Practice) assessment model. Instead of developing a new BPM maturity model, an existing BPM maturity model was evaluated according to its use through several phases. Findings are thus based on what practitioners experienced as difficult or important during the application of a generic maturity model, and gradual adjustments in the PoP assessment model were made to observe and validate situational needs. Another option could be using the same BPM maturity model in case studies. For instance, the BPM Capability Framework [20] or the Process and Enterprise Maturity Model [11] are two examples among the leading BPM maturity models in the literature. However, since empirical evidence for a generic BPM maturity model is generally lacking [36], the practical approach of co-tailoring a PoP assessment model to uncover what is important in practice can be motivated.

3.1 Data Collection Phase

The first phase of data collection was bipartite. First, in a pre-study [38], data were collected from eight practitioners, through observing the challenges faced when they used a generic BPM maturity model [9]. The maturity model was selected by one of the practitioners, based on its availability as a questionnaire and survey tool (compared to traditional research papers), as well as on its explicit recognition that the model is based on practice and developed for practitioners as a target group. The practitioners were also asked for suggestions on what to evaluate (i.e. the most relevant measures in the maturity model).

In addition, data were collected during a university course on BPO between 2010 and 2014, for which an assignment was conducted in the respective organizations of the participants. The assignment was based on the process staircase (of Sect. 2) for evaluating their BPM work practice and the effects of BPO. The practitioners had a common base in BPO knowledge (according to the course) and different BPM experience in their organizations. The assignment was directed to BPM tasks and effects experienced by practitioners in their roles as quality manager, business developer or system developer. Twenty-two course participants from the public sector, eleven from industry and three from non-profit organizations suggested possible measurement items in 117 questions. The sampling of these 36 practitioners was random based on course participation and participation in the assignment.

As such, we obtained the baseline of a tailored BPM maturity model (i.e. PoP assessment model version 1) based on the pre-study, supplemented by suggestions from practitioners in 21 different organizations.

3.2 Validation and Testing Phase

The second phase of validation was based on seven practitioners, namely four from the public sector (Municipality A) and three from industry (Company AB). These practitioners took part in a workshop to evaluate the findings presented in a 'workshop template' (i.e. PoP assessment model version 1). During the four-hour workshop, the template was used in two mixed groups to enable knowledge sharing and further analysis. Questions in each step of the 'process staircase' and areas of interest were coded by colors to reach the total number of practitioners who considered the questions to be important (green), questions not selected (red) and questions that can be merged into one question (yellow). In total, 117 questions were suggested by dividing questions into several issues, merging multiple issues into one, adding questions and changing the terminology of question formulations. The workshop was closed with a discussion on the meaning of an appropriate maturity model design. The second version of the BPM maturity model was developed based on content and design issues, e.g. structure, headings, subtitles of area of interests, questions and measures (i.e. PoP assessment model version 2). The next step in the validation phase was follow-up interviews based on the BPM maturity model, validated by five practitioners. These respondents were randomly selected based on an open invitation in a national BPM network. Each follow-up interview (by phone call) was accompanied by an email with the BPM maturity model (i.e. PoP assessment model version 2). During a 30-minute conversation, challenges, problems or possibilities in use were identified, as well as content and design issues, resulting in a further developed BPM maturity model with a survey tool (i.e. PoP assessment model version 3).

The third phase (testing) was based on seventeen practitioners from the public sector (i.e. IT department in Municipality B), using the assessment survey of the BPM maturity model (i.e. PoP assessment model version 3). The call for using the BPM maturity model survey was made as an open invitation in a national BPM workshop.

4   Results

Table 1 summarizes the lessons learned from our empirical study, and relates them to the challenges identified in the literature (Sect. 2) to re-interpret prior work.

Table 1. Lessons learned from our empirical study, related to the challenges in the literature

Lessons learned (L) related to Challenges (C)
L1: A need to know where to find BPM maturity models (e.g. a particular domain or field, different versions in different channels) – C1
L2: A need for explicit descriptions of possible purposes for using a BPM maturity model (e.g. identifying needs, benchmarking, defining a baseline for improvements, illuminating important aspects) – C3
L3: A need for deeper domain knowledge (BPO and organizational) in order to translate assessment questions/statements into a specific area of interest – C4
L4: A need for an explicit description of scope on width (e.g. BP/BPM/ BPO-level and organization/department/team-level) as well as an explicit scope on depth (e.g. the number and type of business processes) – C3, C4
L5: Open questions to collect insights from practitioners who cannot see questions/statements that fit their need, as well as feedback on the maturity model – C2, C6
L6: A need for questions on BPO knowledge of the respondents, their years of BPM experience and years of working in the organization to evaluate their domain knowledge and be able to validate the result – C2, C8
L7: A need to understand the rationale and value propositions in using a maturity model (e.g. effects on the organization, on business performance/improvements) – C2, C3, C7
L8: A need for a maturity assessment 'order form' to tailor the maturity model to the specific assessment context and purpose of use, as well as to analyze the results in the intended ways – C3, C4, C6, C7, C8
L9: A need for even fewer questions (what to do) and statements (what effects to achieve) when the purpose is only to define a baseline – C4, C6
L10: Multiple management dimensions in a particular maturity model – C5
L11: A need for stronger support to help users interpret the question correctly. Respondents with fairly similar BPO knowledge may understand and interpret the same questions in the BPM maturity model differently – C6
L12: A need to understand what to measure in the different steps/stages in a particular maturity model, and how the outcomes of maturity are measured – C6, C7

The lessons learned in Table 1 relate to the pre-assessment and assessment phase of a maturity model. For instance, users need information about maturity models to understand and select a particular maturity model (L1), as well as information on how to use it during the assessment and analysis (L2). Furthermore, respondents need a design to understand the meaning of the assessment questions or statements to answer them as intended (L5, L9, L11). Nonetheless, it remains difficult to handle the multiple levels of BPM knowledge for diverse users and respondents, as well as different scopes in the same maturity assessment. Hence, another task in tailoring might be targeting different target groups and contexts, i.e. to design 'maturity assessment pieces of the puzzle': *'If an assessment model is used in its entirety for both managers and employees, it is probably difficult to get a good result. You should probably use some selected parts in the model for the employees, and compare the responses of management and employees.'* Furthermore, Table 1 shows that some of the lessons learned represent multiple management dimensions (L10), with implications on both the maturity model and the organization. For instance, while the maturity model under study addressed the scope of business process modeling, the sub-titles of areas of interest might mislead users and respondents towards BPO, and thus leading to vague questions. As a result, multiple management dimensions occurred in the maturity assessment. Another example of an area of interest was 'motivation', for which the assessment questions can evaluate (1) business processes (e.g. employee motivation for change), (2) business process descriptions (e.g. motivation within the modeling team and the model owner to publish process descriptions), (3) as well as the BPM work practice in the organization (e.g. to enable motivation for BPO). Multiple dimensions can occur according to the scope in maturity models (i.e. BPO maturity, BPM maturity, process maturity or modeling maturity) and the scope in organizations (i.e. the geographical level of an assessment or area of interest) (L4). In addition, a relation exists between the respondents and their domain knowledge to answer the assessment questions (L6), as well as between the maturity model users and their knowledge to set the pre-conditions and translate the assessment questions or statements (L3). User ability and skills are needed to conduct a BPM maturity assessment. Users are in charge of finding and 'translating' the maturity model, and are not always the same persons as the respondents. They typically are the process owners or process managers. Respondents might be the maturity model users, but also other employees. Maturity models can be more supportive for practitioners by providing information on how to conduct the assessment (e.g. forms or sheets for taking surveys) and how to evaluate the results (e.g. audits, revisions, feedback to respondents and target groups).

In addition, since BPO is critical to managers, metrics need to be identified according to the purpose and intentions of BPO and its desired effects (L12). Accordingly, a maturity model should include recommendations with explanations and advice focusing on how to measure the outcomes of maturity in order to know how and when to move up the so-called 'process stairs' or maturity levels. Deciding on a business strategy and the related organizational goals involves an analysis (e.g. regarding market demands or regulations, customer satisfaction, costs versus benefits) with implications for motivating the next step to be taken. In order to know when an organization is "ready" to move to the next step, alignment between its business strategy and its assessment results can be seen as one way to evaluate the efforts and results. Based on the BPO foundation, the intended degree and desirable effects of BPM will motivate employees to take the next step in each business case (L7). Given the lack of a single measurement system that satisfies all stakeholders, an organization may tailor how to measure the outcomes of BPM maturity to know when and how to move up.

Finally, the empirical findings stress the need to stipulate specific pre-conditions in order to support the users and respondents using a BPM maturity model. Consequently, a maturity assessment 'order form' (L8) can be used in the pre-assessment phase to support the tailoring of the maturity model to a particular assessment domain (i.e. environment, context and use case). An 'order form' would also facilitate the analysis of the assessment results aligned with business strategies in the intended ways.

5   Discussion

The potential challenges (Sect. 2) and lessons learned (Sect. 4) indicate possible opportunities for tailoring a BPM maturity model. Since they represent model-related and organization-related characteristics, tailoring may impact on both the pre-assessment phase and the assessment phase of using a maturity model. Thus, this paper goes beyond the strict model-related characteristics as investigated in most design studies by adding practical considerations. A *key element for tailoring* is defined as an element or aspect of a maturity assessment that risks being interpreted in diverse ways, and thus can lower the quality of the assessment results. We followed a concept modeling approach to uncover key elements and their relations in a tailoring framework (Fig. 1). Table 2 explains each central element for tailoring.

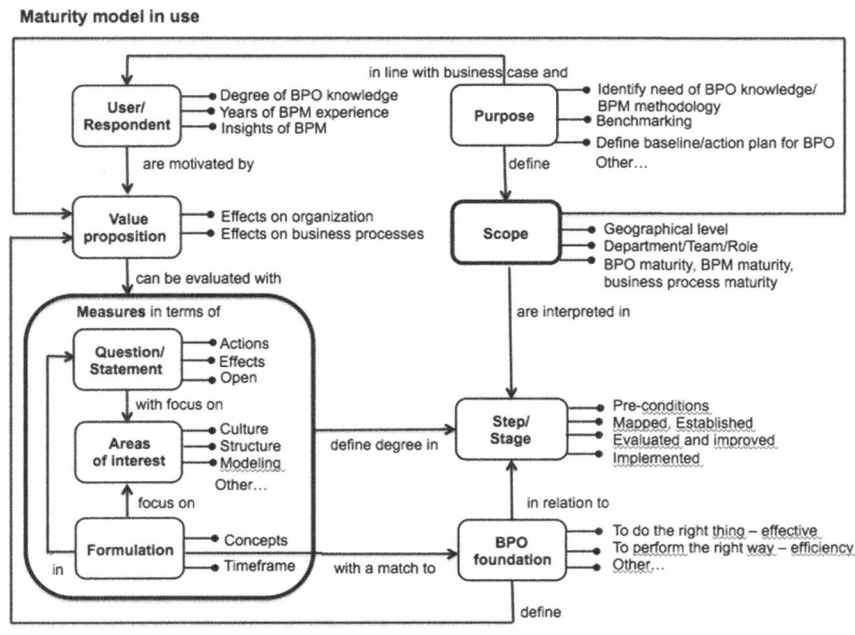

Fig. 1. The tailoring framework for a BPM maturity model in use.

Table 2. Seven central elements with implications for tailoring.

Elements	Findings in literature and practice to take into account
1: Purpose	• As-Is versus To-Be • Identifying the need or possibility to acquire BPO knowledge • Identifying the need for BPM as a methodology • Benchmarking BPM work practice between departments, teams, roles • Benchmarking management versus employees • Obtaining a certification for external recognition • Defining a baseline for BPO, i.e. a roadmap or action plan • Illuminating the expected leadership
2: User/respondent	• Target group • Insights into BPM work practice, years of BPM experience • Degree of BPO knowledge
3: Scope	**For maturity in an organization** • Geographical level (national, regional or local) • Organization/department/team/role • Type of processes/intra- or inter-organizational **For maturity in BPO** • BPO maturity, BPM maturity, business process maturity
4: Value propositions	• Effects on the organization (e.g. leadership, structure, performance) • Effects on business processes (e.g. process performance)
5: Step/stage	**Degree of BPO** • Pre-conditions for BPO (e.g. processes are mapped, established, evaluated, improved and implemented)
6: BPO foundation	• Doing the right things from a customer perspective – effective and innovative • Doing business in the right way – efficiency with resource management in organization flows with IT as enabler and demand-driven use
7: Measures	**Approach** • Questions: actions to take • Statements: effects to achieve • Open questions: feedback to learn from **Areas of interest** • Capability areas (e.g. process modeling, process deployment, process optimization, process governance, a process-oriented culture, a process-oriented structure) **Formulation** • Using appropriate concepts with a match to the respondents • Time frame for the assessment (i.e. As-Is versus To-Be) and in the assessment (e.g. based in one specific initiative or a period with waves of initiatives)

The relations on the right side of Fig. 1 are explained first. Depending on the purpose of using a BPM maturity model, the scope of a particular assessment needs to be defined in order to know the focus to be taken. Also the BPO construct needs to be interpreted in line with the determined purpose of BPM initiatives in a particular organization. A defined BPO construct will eventually affect the content within each step or stage, representing the degree of BPO. The relations on the left side of Fig. 1 mainly focus on the measures to be defined according to appropriate scales, and this to decide on the degree of BPO. Measures can be questions or statements expressed by formulations, which deal with different capability areas or areas of interests, and need grounding in the BPO foundation, i.e. objectives to reach the value propositions. Formulations might relate to different degrees of BPO in an organization, depending on the scope. Finally, the selection of users and respondents depends on the purpose of using a BPM maturity model and their interest in the value propositions.

Table 3. Design propositions (D) to be adopted in a tailoring template.

Organization-related pre-conditions	
User	**D1:** Determine *the user* degree of BPO knowledge and years of BPM experience
Purpose	**D2:** Determine *the purpose of the assessment* and results in an explicit business case or setting with a fit to the target group
Value proposition	**D3:** Determine *expectations of the maturity assessment results* and how to perform the evaluation towards effects on organization and business processes by steps/stages according to the BPO foundation
Respondent	**D4:** Determine *the respondents'* degree of BPO knowledge, years of BPM experience, insights into ongoing BPM in the organization
Model-related content	
Scope	**D5:** Define *the scope* on a geographical level (national, region or local), organizational level (management or operational) and process level (process types and intra- or inter-organizational)
Measurement	**D6:** Determine the *measurement approach*, i.e. questions (actions to take) or statements (effects to achieve) and the need for open questions
Measures: Areas of interest	**D7:** Define *areas of interest or capability areas* depending on the purpose and respondents in the business case
Measures: Formulations	**D8:** Define the *time frame* planned for the assessment and for focus during the assessment **D9:** Use concepts that are familiar in the organization to define the *BPO construct* to be used in the MM **D10:** Define *measurements* appropriate to the respondents based on the BPO foundation and value propositions
Measures: Questions/statements	**D11:** Per measure, focus on one area of interest and define the scope, i.e. timeframe for the initiative (project/continuous improvement) AND with a match to the BPO foundation AND in relation to one step/stage **D12:** Use open questions, e.g. "Is a question missing that should be asked in a BPM maturity assessment?" or "Would you like to collect opinions about the questionnaire, e.g. which questions were difficult?"

From the tailoring framework, we derived twelve design propositions (D) in a 'tailoring template' to be used by the maturity model users and respondents (Table 3). These design propositions for tailoring are divided into *organization-related elements* to determine the pre-conditions for a BPM assessment, and *model-related elements*.

6 Conclusions and Limitations

While many BPM maturity models exist, their adoption in and adaptation to organizations are lagging behind. Although the general ideas of a BPM maturity model are well understood, organizations need to carefully prepare for using a BPM maturity model. If an organization lacks competent BPM managers, respondents can rely on external consultancy or academic support to strengthen their internal knowledge about BPO, their work practice, and to continuously develop that knowledge. In addition, organizations might need the support and guidance of external consultants and/or academic mediators/mentors to define the domain and version of a maturity model, as well as to select respondents and to conduct the assessment (e.g. assessment form, data analysis, etc.). Hence, we introduce *an emergent tailoring framework and template* to facilitate the use of BPM maturity models. Our study builds on previous design ideas [18, 29], and supplements them with evidence, examples and new findings. By using a 'tailoring template', a better fit in the following areas is presumed:

- Awareness of the business context to decide on pre-conditions and timing;
- Awareness and selection of possible purposes and appropriate respondents;
- Explicit scope (i.e. width and depth) and defined BPO constructs;
- Insight into inputs for measuring maturity and how the outcomes are measured;
- Explicit focus and formulation of the assessment questions and statements.

Our main research limitation is that we only used one BPM maturity model, albeit through different phases. Although it concerns a 'non-dominant' model in the process literature, it was seen as practical to use by the practitioners. Thus, a 'bottom-up' approach is used to let 'the practice speak' and a 'dominant' BPM maturity model (i.e. mostly cited by scholars) is not necessarily the same as the most used maturity model in practice. Apparently, a 'research-practice gap' exists as researchers develop maturity models that have not gained widespread acceptance in practice [36]. Further research can examine why such a gap exists. Another limitation is that the study relies on a Swedish business context. Thus, another research avenue is to compare our findings with other countries. In addition, our practitioners had different BPM experiences ranging from one year to more than 10 years of practice. One option is to analyze how our data will respond to categorizations such as 'novice', 'experienced' or 'expert' users. Alternatively, the factor of 'education/training' in BPO can be added. Our study also indicates the need for more knowledge of how practitioners apply BPM maturity models. There is not necessarily a need of more BPM maturity models, but rather for clarifying the issues related to the BPM work practice and the business context [31].

In conclusion, our design propositions may serve as a foundation to refine a meta-model for BPM maturity model design in order to further explore the question: 'What makes a useful maturity model?' [18]. In such a meta-model, context factors

[31] should be considered on each level or dimension of scope (e.g. geographical, organizational and process) to design a context-sensitive maturity model. For example, nationality can be an important context factor to differentiate the content and the application of a maturity model, i.e. based on a different national culture and values. Nonetheless, our 'tailoring template' needs to be used and validated first in a wider range of BPM maturity assessment practices to further strengthen our findings.

References

1. Becker, J., Knackstedt, R., Pöppelbuß, J.: Developing maturity models for IT management. Bus. Inf. Syst. Eng. **1**(3), 213–222 (2009)
2. Buh, B., Andrej Kovačič, A., Štemberger, M.I.: Critical success factors for different stages of business process management adoption – a case study. Econ. Res. **28**(1), 243–258 (2015)
3. Choong, K.K.: Are PMS meeting the measurement needs of BPM? A literature review. BPM J. **19**, 535–574 (2013)
4. Christiansson, M.-T.: Process orientation in inter-organisational co-operation. In: Nilsson, A.G., Pettersson, J.S. (eds.) On Methods for Systems Development in Professional Organisations. Studentlitteratur, Lund (2001)
5. Cronemyr, P., Danielsson, M.: Process management 1-2-3 – a maturity model and diagnostics tool. Total Qual. Manag. Bus. Excell. **24**(7–8), 933–944 (2013)
6. de Bruin, T.: Business process management: theory on progression and maturity. Dissertation, Queensland University of Technology (2009)
7. de Bruin, T., Rosemann, M., Freeze, R., Kulkarni, U.: Understanding the main phases of developing a maturity assessment model. In: ACIS Proceedings (2005)
8. Fettke, P., Zwicker, J., Loos, P.: Business process maturity in public administrations. In: vom Brocke, J., Rosemann, M. (eds.) Handbook on BPM, vol. 2, 2nd edn. Springer, Heidelberg (2015). doi:10.1007/978-3-642-01982-1_18
9. Ganesan, E.: A practitioner's guide to assess the maturity and implementation of enterprise process modeling using CEProM assessment framework. In: BPTrends (2011)
10. Goldkuhl, G.: The research practice of practice research: theorizing and situational inquiry. Syst. Signs Actions **5**, 7–29 (2011)
11. Hammer, M.: The process audit. Harv. Bus. Rev. **85**, 111–123 (2007)
12. Henderson-Sellers, B., Ralyté, J., Ågerfalk, P.J., Rossi, M.: Tailoring a constructed method. In: Situational Method Engineering, pp. 169–194. Springer, Heidelberg (2014). doi:10.1007/978-3-642-41467-1_7
13. Hevner, A.R., March, S.T., Park, J., Ram, S.: Design science in information systems research. MIS Q. **28**, 75–105 (2004)
14. Indulska, M., Recker, J., Rosemann, M., Green, P.: Business process modeling: current issues and future challenges. In: Eck, P., Gordijn, J., Wieringa, R. (eds.) CAiSE 2009. LNCS, vol. 5565, pp. 501–514. Springer, Heidelberg (2009). doi:10.1007/978-3-642-02144-2_39
15. Kohlbacher, M.: The effects of process orientation: a literature review. BPM J. **16**(1), 135–152 (2010)
16. Kohlbacher, M., Reijers, H.A.: The effects of process-oriented organizational design on firm performance. BPM J. **19**(2), 245–262 (2013)
17. Mettler, T., Rohner, P.: Situational maturity models as instrumental artifacts for organizational design. In: DESRIST Proceedings (2009)

18. Pöppelbuß, J., Röglinger, M.: What makes a useful maturity model? In: ECIS Proceedings, paper 28 (2011)
19. Rohloff, M.: Case study and maturity model for business process management implementation. In: Dayal, U., Eder, J., Koehler, J., Reijers, H.A. (eds.) BPM 2009. LNCS, vol. 5701, pp. 128–142. Springer, Heidelberg (2009). doi:10.1007/978-3-642-03848-8_10
20. Rosemann, M., de Bruin, T.: Application of a holistic model for determining BPM maturity. In: BPTrends (2005)
21. Rosemann, M., vom Brocke, J.: The six core elements of business process management. In: vom Brocke, J., Rosemann, M. (eds.) Handbook on BPM, vol. 1, 2nd edn., pp. 109–124. Springer, Heidelberg (2015). doi:10.1007/978-3-642-00416-2_5
22. Röglinger, M., Pöppelbuß, J., Becker, J.: Maturity models in business process management. BPM J. 18, 328–346 (2012)
23. Skålén, P., Gummerus, J., von Koskull, C., Magnusson, P.R.: Exploring value propositions and service innovation: a service-dominant logic study. J. Acad. Mark. Sci. 43(2), 137–158 (2015)
24. Spanyi, A.: Beyond process maturity to process competence. In: BPTrends (2004)
25. Tarhan, A., Turetken, O., Reijers, H.: Business process maturity models: a systematic literature review. Inf. Softw. Technol. 75, 122–134 (2016)
26. Tarhan, A., Turetken, O., Reijers, H.: Do mature business processes lead to improved performance? - A review of literature for empirical evidence. In: ECIS Proceedings, paper 178 (2015)
27. Van Looy, A.: Business Process Maturity. SBPM. Springer, Cham (2014). doi:10.1007/978-3-319-04202-2
28. Van Looy, A., De Backer, M., Poels, G.: A conceptual framework and classification of capability areas for business process maturity. Enterp. Inf. Syst. 8(2), 188–224 (2014)
29. Van Looy, A., De Backer, M., Poels, G., Snoeck, M.: Choosing the right business process maturity model. Inf. Manag. 50(7), 466–488 (2013)
30. Van Looy, A., De Backer, M., Poels, G.: Defining business process maturity, a journey towards excellence. Total Qual. Manag. Bus. Excell. 22(11), 1119–1137 (2011)
31. vom Brocke, J., Zelt, S., Schmiedel, T.: On the role of context in business process management. Int. J. Inf. Manag. 36(3), 486–495 (2016)
32. vom Brocke, J., Schmiedel, T., Recker, J., Trkman, P., Mertens, W., Viaene, S.: Ten principles of good business process management. BPM J. 20, 530–548 (2014)
33. vom Brocke, J., Sinnl, T.: Culture in business process management: a literature review. BPM J. 17, 357–377 (2011)
34. Walsham, G.: Doing interpretive research. Eur. J. Inf. Syst. 15, 320–330 (2006)
35. Wendler, R.: The maturity of maturity model research: a systematic mapping study. Inf. Softw. Technol. 54, 1317–1339 (2012)
36. Wolf, C., Harmon, P.: The state of BPM. In: BPTrends (2014)
37. zur Muehlen, M.: Class notes: BPM research and education. In: BPTrends (2008)
38. Christiansson, M.-T., Granström, K.: Sharpening the knowledge domain transfer in practice research design - a case study. Syst. Signs Actions 6(1), 22–45 (2012)

A New Framework for Defining Realistic SLAs: An Evidence-Based Approach

Minsu Cho[1,2], Minseok Song[2(✉)], Carlos Müller[3], Pablo Fernandez[3], Adela del-Río-Ortega[3], Manuel Resinas[3], and Antonio Ruiz-Cortés[3]

[1] Ulsan National Institute of Science and Technology, Ulsan, Korea
mcho@unist.ac.kr
[2] Pohang University of Science and Technology, Pohang, Korea
mssong@postech.ac.kr
[3] University of Seville, Seville, Spain
{cmuller,pablofm,adeladelrio,resinas,aruiz}@us.es

Abstract. In a changing and competitive business world, business processes are at the heart of modern organizations. In some cases, service level agreements (SLAs) are used to regulate how these business processes are provided. This is usually the case when the business process is outsourced, and some guarantees about how the outsourcing service is provided are required. Although some work has been done concerning the structure of SLAs for business processes, the definition of service level objectives (SLOs) remains a manual task performed by experts based on their previous knowledge and intuition. Therefore, an evidence-based approach that curtails humans involvement is required for the definition of realistic while challenging SLOs. This is the purpose of this paper, where performance-focused process mining, goal programming optimization techniques, and simulation techniques have been availed to implement an evidence-based framework for the definition of SLAs. Furthermore, the applicability of the proposed framework has been evaluated in a case study carried out in a hospital scenario.

Keywords: Service level agreement · Process mining · Process performance indicators · Optimization · Goal programming · Simulation

1 Introduction

In a changing and competitive business world, business processes are at the heart of modern organizations [1]. In some cases, service level agreements (SLAs) are

This work was partially supported by the European Commission (FEDER), the European Union Horizon 2020 research and innovation programme under the Marie Sklodowska-Curie grant agreement No. 645751 (RISE_BPM), the Spanish and the Andalusian R&D&I programs (grants TIN2015-70560-R, P12-TIC-1867), the National Research Foundation of Korea (No. NRF-2014K1A3A7A030737007).

© Springer International Publishing AG 2017
J. Carmona et al. (Eds.): BPM Forum 2017, LNBIP 297, pp. 19–35, 2017.
DOI: 10.1007/978-3-319-65015-9_2

used to regulate how these business processes are provided. This is usually the case when the business process is outsourced, and some guarantees about how the outsourcing service is provided are required [2].

Although some work has been done concerning the structure of SLAs for business processes [2], the problem of defining the actual service level objectives (SLOs), which are essential factors of an SLA denoting requirements on the service performance, in a specific business process is largely unaddressed. This issue involves first choosing the process performance indicators (PPIs) that should be considered in the SLA and, second, defining their desired target. This target must be challenging, but achievable to ensure a good process performance. A consequence of this lack of methodology for defining SLOs is that, in the current state of practice, the definition of SLOs is usually carried out by experts based on their previous knowledge and intuition, and sometimes following a trial and error model. This is far from desired, since, according to [3], definition of objectives requires a theory and a practical base and it should meet certain requirements: not being based on experts opinion, but on measurement data; respecting the statistical properties of the measure, such as measure scale and distribution, and be resilient against outlier values; and being repeatable, transparent and easy to carry out.

To overcome this problem, in this paper, we propose a framework that includes a series of steps for defining SLAs with a systematic evidence-driven approach. The proposed method covers the understanding of current behaviors of business processes, defining SLOs, deriving optimized SLOs with improvement actions, and evaluating the expected effects with a simulation. Specifically, in this paper, we present a proposal to implement the first three steps. This proposal supports a broad range of PPIs and employs performance-focused process mining, optimization techniques for multi-objective programming, and simulation techniques. The contributions of our research are as follows: (i) connecting the realms of process mining and SLAs; and (ii) proposing a new systematic approach to defining SLAs based on evidence. The applicability of our approach has been demonstrated with an experimental evaluation based on a hospital scenario.

The remainder of this paper is organized as follows. Section 2 introduces our evidence-based framework for the definition of SLAs. Then, in Sect. 3, we describe the first three steps in a formal way: (i) how to derive PPIs and current SLOs (CSLOs) in Sect. 3.1; and (ii) how to optimize SLOs in Sect. 3.2. Section 4 describes the effectiveness of our approach with the experimental evaluations. The summarized related works are presented in Sect. 5, and finally, Sect. 6 concludes the work and describes future directions.

2   An SLA Definition Framework Using an Evidence-Based Approach

In this section, we introduce the SLA definition framework using an evidence-based approach. As depicted in Fig. 1, the framework consists of 6 steps:

(1) process mining analysis, (2) SLOs inference, (3) SLOs optimization, (4) simulation analysis, (5) evaluation, and (6) SLA definition. Initially, we conduct process mining analysis to calculate PPIs using event logs extracted from information systems. On the basis of PPIs, results from process mining analysis, CSLOs are inferred. These CSLOs represent the current behavior for PPIs. After that, CSLOs are used in the SLOs optimization step that generates desired service level objectives (DSLOs) by applying some optimization techniques. The next step is to build a simulation model and conduct a simulation analysis for a scenario based on the optimization. Then, the evaluation step is performed to analyze the deviation between PPIs from the simulation and the DSLOs. In such a step, according to the evaluation result, the process can revert to either *Step 2* or *Step 3*, in the other case, it can proceed to *Step 6*. Specifically, the *SLOs inference activity* is performed in the state of a high deviation between the PPIs and DSLOs; the *SLOs optimization activity* is executed in case of a low deviation between two values. Finally, new SLAs are derived based on the DSLOs that successfully pass the evaluation step.

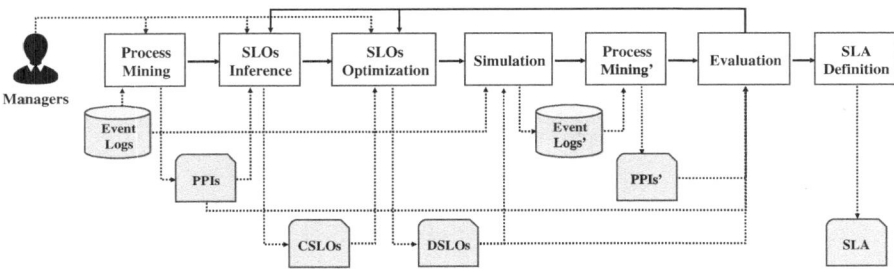

Fig. 1. Overviews of the proposed framework

As already mentioned, in this paper we focus on *Steps 1, 2, and 3* as a first approach towards supporting the whole SLA definition framework using an evidence-based approach. In order to exemplify these steps, from a user interaction perspective, Fig. 2 depicts a sequence of four mockups describing the expected interaction flow of a manager using the system in a given scenario. Specifically, the first mockup corresponds with a particular PPI selection over the outcomes of the process mining analysis (*Step 1*). Once the user selects the subset of PPIs to be optimized, the second mockup presents the essential step where current SLOs are spotted for each PPI (*Step 2*); based on the business goals, in this point, the manager can specify a desired SLO and check the appropriate potential actions to achieve the expected SLO joint with an estimated impact of the actions (as a starting point, the system calculates an estimation based on the current data that can then be tuned by the manager). Next, in the third mockup, a global set of constraints can be established typically including costs over the improvements. Finally, in the fourth mockup, the result of the optimization (*Step 3*) is shown describing the proposed improvement actions along the desired SLO and the expected metrics according to the global constraints.

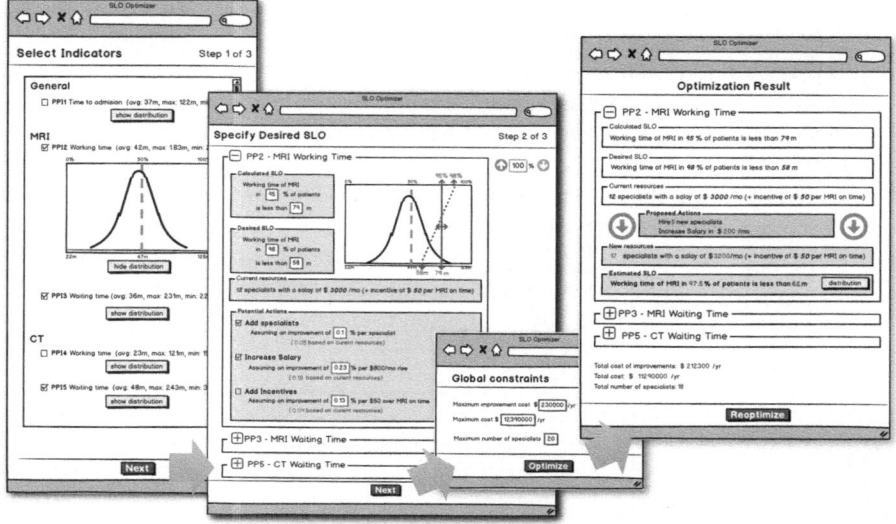

Fig. 2. User interaction flow

3   A Proposal for Obtaining DSLOs

In this section, we detail a proposal for the first three steps of the framework, which is the focus of this paper. The proposal supports a broad range of PPIs and uses a goal programming approach as the optimization technique for the SLOs optimization step.

3.1   Process Mining Analysis and SLOs Inference

Among the different perspectives involved in process mining [4], our first step focuses on the performance perspective and tries to infer the performance of a current process from its past executions stored in an event log. In such a step, a set of pre-defined PPIs (i.e., PPIs catalog) is applied, and PPIs are computed from the event log. After that, SLOs are inferred based on the calculated PPIs. In contrast to the manual approach currently followed to define SLOs, this paper proposes an evidence-based approach as an alternative. Therefore, in the SLOs inference step, managers only have a decision to select target PPIs because all PPIs are not the key performances for a process. In other words, a couple of

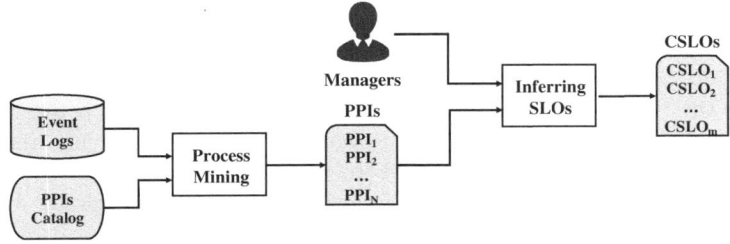

Fig. 3. Process mining analysis & inferring SLOs

principal PPIs are selected and inferred to CSLOs as targets to be improved. Figure 3 provides the steps for process mining analysis and SLOs inference.

We now give a detailed explanation with formal definitions for each part. Event logs, which are the inputs of process mining, are a collection of cases, where a case is a sequence of events (describing a trace). In other words, each event belongs to a single case. Events have four properties: activity, originator, event type, and timestamp. Thus, events can be expressed as assigned values for these four properties. These are defined as follows.

Definition 1 (Events, Cases, and Event Log). *Let A, O, ET, T be the universe of activities, originators, event types, timestamps, respectively. Let $\mathcal{E} = A \times O \times ET \times T$ be the universe of events. Note that events are characterized by properties. For each event e and each property p, $p(e)$ denotes the value of property p for event e (e.g., $act(e)$, $type(e)$, $time(e)$ are the activity, the event type, and the timestamp of the event e, respectively). If there is no assigned value of the event e to the property p, we use $\perp$. Let $C = E^*$ be the set of possible event sequences (i.e., cases). An event log $\mathcal{L} \in \mathbb{B}(C)$ is the set of all possible multi-sets over C.*

A simple example log is provided in Table 1. In the table, 24 events for four cases are included, and each line corresponds to a trace represented as a sequence of activities. For example, the trace of the case 1 refers to a process instance where A was started by Paul at 09:00 and completed at 10:00, B was started by Mike at 10:20 and completed at 12:00, and C was started by Allen at 13:00 and completed at 13:30. Also, event IDs are determined by the order of cases and timestamps of events (i.e., E1: $A_{Start}^{Paul,09:00}$, E2: $A_{Complete}^{Paul,10:00}$, ..., and E24: $D_{Complete}^{Allen,17:30}$). This log will be used as a running example.

Table 1. Running example log

Case	Trace
Case 1	$<A_{Start}^{Paul,09:00}, A_{Complete}^{Paul,10:00}, B_{Start}^{Mike,10:20}, B_{Complete}^{Mike,12:00}, D_{Start}^{Allen,13:00}, D_{Complete}^{Allen,13:30}>$
Case 2	$<A_{Start}^{Paul,10:30}, A_{Complete}^{Paul,11:00}, C_{Start}^{Chris,12:10}, C_{Complete}^{Chris,13:00}, D_{Start}^{Allen,14:00}, D_{Complete}^{Allen,15:00}>$
Case 3	$<A_{Start}^{Paul,12:00}, A_{Complete}^{Paul,12:30}, B_{Start}^{Mike,14:00}, B_{Complete}^{Mike,15:00}, D_{Start}^{Allen,15:30}, D_{Complete}^{Allen,16:30}>$
Case 4	$<A_{Start}^{Paul,13:00}, A_{Complete}^{Paul,14:00}, C_{Start}^{Chris,14:30}, C_{Complete}^{Chris,15:30}, D_{Start}^{Allen,16:00}, D_{Complete}^{Allen,17:30}>$

Based on the event log, we identify events to be used for calculating PPIs through two elements: entity types and entity identifiers. The entity type includes activity and originator. The entity identifiers signify the possible values that belong to the entity type. For example, in Table 1, A, B, C, and D are the entity identifiers of the entity type activity. Based on these two elements and the log (i.e., the event log, the entity type, and the entity identifier), required events are filtered and extracted through the ψ function. After that, extracted events are calculated based on measures such as count, working time, and waiting time. A PPI $(P_n(M(\mathcal{E})))$ is defined as calculating n-th percentile (P_n) from computed measure values for the filtered events. The PPI is defined as follows.

Definition 2 (Process Performance Indicators). *Let T and V be the universe of entity types and universe of possible values, respectively. For each entity type $t \in T$, V_t denotes the set of possible values, i.e., the set of entity identifiers of type t. Let $\psi \in L \times T \times V_T \Rightarrow \mathcal{E}$ is a function that finds out the set of events from an event log for a given entity type and an entity identifier (where, $\mathcal{E}$ is the set of events). M is the measures such as count, working time, waiting time, duration, etc. $P_n(M(\mathcal{E}))$ is the process performance indicator from an event log for a given entity type and an entity identifier, and a measure (where, $P_n =$ be the n-th percentile function). Note that P_{25}, P_{50}, P_{75} are 1st quantile, median, 3rd quantile, respectively.*

For example, we can get following examples from the Table 1; $\psi(L, Activity, A)$ = {E1, E2, E7, E8, E13, E14, E19, E20}, $\psi(L, Originator, Allen)$ = {E5, E6, E11, E12, E17, E18, E23, E24}. As an example of PPIs, the median of working time for $\psi(L, Activity, A)$ is calculated as 30 min from {60, 30, 30, 30}, and it also can be denoted as follows: *median of working time of the A is 30* min.

The next step is to infer SLOs based on the calculated PPIs using process mining. SLO is defined as follows.

Definition 3 (Service Level Objectives and Inferring function). *Let M be the universe of measurements. x is the target value of the measurement m, and $P(t)$ is the function deriving probability of t. A SLO $P(m \leq x) \geq n\%$ is the probability(m) that measurement is less than x must be at least $n\%$. Let $I \in \Gamma(P_x(M(\mathcal{E}))) \Rightarrow \{P(m \leq x) \geq n\%\}$ be a function that infers the SLO from the PPI.*

SLO is defined as *a probability that a measure of cases that have the entity identifier $\leq$ value must be more than $n\%$*. CSLOs are automatically inferred from PPIs using the I function. Overall structures of CSLOs and PPIs are quite similar; thus, we can easily establish CSLOs using given PPIs. As we explained earlier, PPI is defined as *n-th percentile* of *a measure* of *an entity identifier* is *value*. Based on the PPI, CSLO becomes *a measure* of *an entity identifier* must be less than the *value* in *n%* of cases. For example, in Table 1, one of the PPIs, the *median (50th percentile)* of *working time* of the *activity A* is *30* min (i.e. PPI_1). Then, the related CSLO becomes the *working time* of the *activity A* must be less than *30* min in *50%* of the cases. Also, there is another PPI that the *median* of *working time* of the originator *Allen* is *60* min (i.e., PPI_2). Then, the corresponding CSLO becomes the *working time* performed by *Allen* must be less than *60 minutes* in *50%* of the cases.

3.2 SLOs Optimization

The objective for the optimization step is to maximize the whole effect by minimizing the target value of each calculated SLO while maintaining it achievable and realistic by selecting the best improvement actions that enhance the process performance. Therefore, it needs a multi-objective programming approach to accomplish multiple goals. We employ the goal programming (GP) approach [5]. The goal programming method is one of the popular approaches for the multi-objective programming problem [5]. Figure 4 shows the SLOs optimization step. In our approach, the inputs of the GP model are *improvement actions, CSLOs*, and *business constraints*. We assume that improvement actions are given based on prior knowledge or qualitative research (e.g., interviews and surveys). Employing more resources and providing incentives are a part of the typical examples of the actions. As explained in Sect. 3.1, CSLOs are derived from event logs. Finally, a manager has to determine demands and constraints including costs of implementation actions, expected SLOs and importance of each SLO. Here, the expected SLOs signify manager's expectation regarding the derived SLOs. On the basis of three inputs, a GP model is constructed, and the output of the model are how many and what improvement actions are used for each goal and the minimized SLOs (DSLOs).

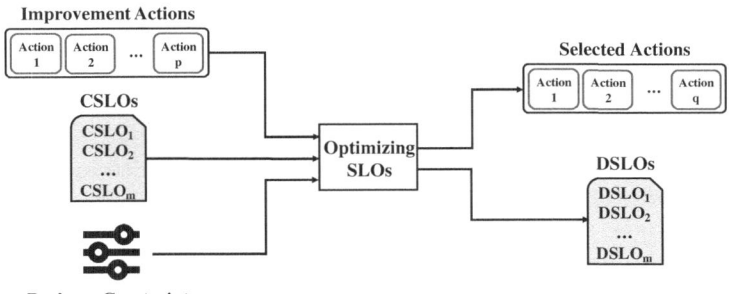

Fig. 4. Optimizing SLOs

Before explaining the GP model, we introduce the symbols that are described in Table 2.

$P_n(\mu, \sigma^2)$ denotes the percentile function for a normal distribution with mean (μ) and variance (σ^2). In general, the percentile function is defined as the infimum function of the cumulative distribution function [6]. Here, based on two aspects, we consider that percentiles are represented by a normal curve plot and

Table 2. Optimization symbols

Symbol	Meaning
V	Number of entity identifiers in all CSLOs
M	Number of available improvement actions
i	Indices of entity identifiers, $(i = 1, 2, ..., V)$
j	Indices of available actions, $(j = 1, 2, ..., M)$
m	Types of measures, $(m = \{d : duration, wo : working, wa : waiting\})$
$x_{i,j}$	Number of applications of action j for entity identifier i
$l_{i,j}$	Lower bound of number of applications of action j for entity identifier i
$u_{i,j}$	Upper bound of number of applications of action j for entity identifier i
μ_i^m	Current mean of measure m for entity identifier i
σ_i^m	Current standard deviation of measure m for entity identifier i
$f_{i,j}^m$	Effect on mean of measure m of action j for entity identifier i
$h_{i,j}^m$	Effect on std. dev. of measure m of action j for entity identifier i
$c_{i,j}$	Unit cost of method j for entity identifier i
C	Planned implement action cost
X	Target percentage by manager $(0 \leq X \leq 1)$
T	Target value by manager
W	Determined range weight for target value $(0 \leq W \leq 1)$
w_k	Importance of SLO k $(k = 1, 2, ...K)$
$P_n(\mu, \sigma^2)$	n-th Percentile function with μ and σ^2

can be expressed with two variables μ and σ^2. First, there is a principle that large populations follow a normal distribution [7]. Second, the improvement actions in this paper have an effect on decreasing mean and standard deviation of distributions. Figure 5 provides the graphical explanation. In a current distribution for an SLO, the target value based on 95% is V_1. If an improvement action makes the mean decrease without any other changes, the distribution moves to the left. As such, the reduced new target value (V_2) is derived as provided in the left graph of Fig. 5. On the other hand, if an improvement action affects the decrease of the standard deviation, the distribution becomes more centralized than before, and the new target value (V_3) is derived as shown in the middle of Fig. 5. Furthermore, an improvement action can affect to reduction of both mean and standard deviation. Then, as shown in the right of Fig. 5, the target value is decreased as V_5 depending on the decrease of mean and standard deviation. The following is the formalization of the percentile function with the normal distribution.

Definition 4 (Percentile Function). *Let f be the probability density function, the cumulative distribution function F as follows:*

$$F(x) = \int_{-\infty}^{x} f(t)\mathrm{d}t \quad (where, -\infty \leq x \leq \infty)$$

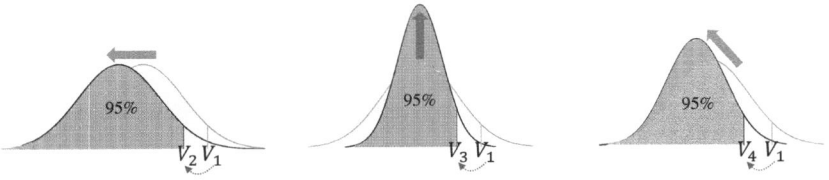

Fig. 5. Effects on target values for SLOs based on improvement actions

With reference to the function F, percentile function is

$$P(p) = inf\{x \leq \mathbb{R} : p \leq F(x)\} \quad (where, \; inf = infimum \; function)$$

for a probability $0 \leq p \leq 1$. Based on the principle that large populations follow a normal distribution, percentile function becomes the inverse function of the cumulative normal distribution function. The cumulative distribution function for normal distribution with μ and σ^2 is as follows.

$$F_x(\mu, \sigma^2) = \frac{1}{2}[1 + erf(\frac{x - \mu}{\sigma\sqrt{2}})] \quad (where, \; erf = error \; function)$$

Let n-th percentile function for normal distribution with μ and σ^2 be defined as follows.

$$P_n(\mu, \sigma^2) = F_x^{-1}(\mu, \sigma^2) = \mu + \sigma\sqrt{2}erf^{-}1(2p - 1)(n\% = p)$$

As we explained earlier, the GP model aims at minimizing the target values of all SLOs by employing improvement actions. Therefore, an individual optimization model for each SLO is constructed. Then the GP model is formulated by combining all optimization models together. An optimization model for each goal is formalized as follows.

Definition 5 (Optimization Model for Each Goal)

O.F.	$DSLO_i = min \; P_X(\mu_i^{m\prime}, \sigma_i^{m\prime 2})$
where,	$\mu_i^{m\prime} = \mu_i^m + \sum_{j=1}^{M} x_{i,j} f_{i,j}^m$
	$\sigma_i^{m\prime} = \sigma_i^m + \sum_{j=1}^{M} x_{i,j} h_{i,j}^m$

Constraints

$$T \times (1 - W) \leq DSLO_i \leq P_X(\mu_i^m, \sigma_i^{m2})$$
$$0 \leq x_{i,1}, x_{i,2}, \ldots, x_{i,M}$$
$$x_{i,1}, x_{i,2}, \ldots, x_{i,M} = integer$$
$$l_{i,j} \leq x_{i,j} \leq u_{i,j} \quad (for \; j = 1, 2, \ldots, M)$$
$$\sum_{i=1}^{V} \sum_{j=1}^{M} x_{i,j} c_{i,j} \leq C$$

As we explained earlier, improvement actions can influence the mean and standard deviation of the distribution for SLOs. As such, the objective function is formalized aiming to minimize the percentile function considering the modified mean and the standard deviation $(P_X(\mu_i^{m\prime}, \sigma_i^{m\prime 2}))$. Here, the updated mean $(\mu_i^{m\prime})$ and standard deviation $(\sigma_i^{m\prime})$ are described as the difference between the current values $(\mu_i^m$ and $\sigma_i^{m2})$ and the effects of the applying improvement actions (i.e., $\sum_{j=1}^{M} x_{i,j} f_{i,j}^m$ and $\sum_{j=1}^{M} x_{i,j} h_{i,j}^m$ that denote the reduction of mean and standard deviation, respectively).

For the constraints in the optimization model, the expected SLO determined by managers is included as a target value with a specific target percentage. Considering the pre-determined expected SLOs, we set the range of $DSLO_i$ that it should be less than or equal to the current value $(P_X(\mu_i^m, \sigma_i^{m2}))$ and greater than or equal to the value from the target value (T) and range weight (W). Moreover, another constraint is that the number of applications for each action $(x_{i,j})$ should be bigger than 0 and integer. In this regard, we can also determine a lower bound $(l_{i,j})$ and an upper bound $(u_{i,j})$ of the number of applications for each action. Furthermore, the cost-related constraint is also included so that total used cost for implementation $(\sum_{i=1}^{V} \sum_{j=1}^{M} x_{i,j} c_{i,j})$ is less than the planned implement action cost (C).

At last, we describe how to formalize the GP model that combines the optimization model for the selected SLOs. The objective function of the GP model considers both the changes of SLOs (i.e., the difference between CSLOs and the minimized SLOs $(DSLO_s)$) and the importance of each goal determined by a manager. Also, constraints and bounds in optimization models for goals are included. Formalization for the GP model is as follows.

Definition 6 (GP Model)

O.F. $max\ Z = w_1 \frac{CSLO_1 - DSLO_1}{CSLO_1} + w_2 \frac{CSLO_2 - DSLO_2}{CSLO_2} + \dots$
               $+ w_K \frac{CSLO_K - DSLO_K}{CSLO_K}$

subject to   *Constraints and bounds in optimization models for goals*

4   Experimental Evaluation

To demonstrate the effectiveness of our proposed approach, we apply it to an examination process in an outpatient clinic and the corresponding log utilized in [8]. In Sect. 4.1, we introduce the examination process and the corresponding log applied in the evaluation. In Sect. 4.2, we describe the results of PPIs calculation and CSLOs conversion. Section 4.3 introduces the setup for the optimization, while Sect. 4.4 provides the results of optimization, i.e., DSLOs.

4.1   Experiment Design and Data Set

As we introduced earlier, we used the examination flows in the outpatient clinic and the corresponding event log. Figure 6 provides the graphical description of the examination process. In the process, patients (i.e., cases) firstly visit a hospital and get both the lab test and the X-ray test. Then, if needed, patients get the electrocardiogram test (ECG). After that, they visit the hospital again and get either computerized tomography (CT) or magnetic resonance imaging (MRI) according to the results of the tests in the first visit. Lastly, the process is finished with the third visit of the patients. The proposed framework was applied to the corresponding log of the examination process. The log included 7000 events performed by 17 resources for 1000 cases.

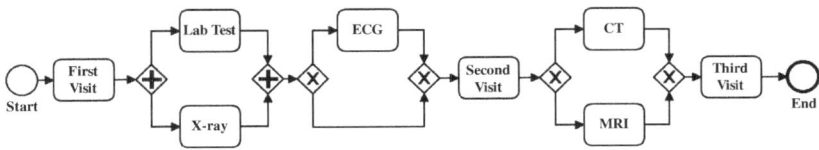

Fig. 6. The examination process used in the evaluation

In the case study, we focused on PPIs defined for the working and waiting time of the test-related activities included in the process. Also, for each indicator, we applied various aggregation functions such as median, first quartile (*1st Q*), third quartile (*3rd Q*), five percentiles (*5%*), and 95 percentiles (*95%*) to understand the distribution of the indicator. We computed PPIs with the examination event log, and Table 3 provides the results in detail.

Table 3. Calculated results of PPIs

(measure: min.)						
Time value	Activity	Median	1st Q.	3rd Q.	5%	95%
Working time	X-ray	20.0	19.0	21.0	17.0	23.0
	Lab Test	20.0	19.0	21.0	17.0	23.0
	ECG	30.0	27.0	33.0	22.0	38.0
	MRI	61.0	56.0	64.3	50.0	71.0
	CT	45.0	44.0	46.0	42.0	48.0
Waiting time	X-ray	30.0	27.0	33.0	22.0	38.0
	Lab Test	30.0	26.7	33.0	22.0	38.0
	ECG	0.0	0.0	0.0	0.0	1.0
	MRI	7223.5	6931.8	7547.2	6478.5	7975.7
	CT	4314.5	3994.7	4651.2	3506.9	5089.0

4.2 Results for PPIs and CSLOs

As described in the Table 3, we identified that MRI had higher working time than any other activities (e.g., the median of working time of MRI was 61 min).

With regard to the waiting time, a couple of activities had higher values than others: MRI and CT. These results were used to determine the candidates for optimization (i.e., CSLOs).

To decide what PPIs are taken into account for the CSLOs extraction, we can consider two types of criteria. First, the indicators that are linked to a critical part in a process, e.g., a primary activity or a sub-process can be selected because they are necessary to improve the process. However, this approach has to be determined by a manager of an organization. In other words, it is required to have a domain knowledge of the process. The other approach is to select problematic indicators that have a high potential to be improved such as indicators that have high volatility or unexpectedly low values. Since that information from the manager was not available in our case study, we selected the second option. Among several PPIs, we selected three of them, and the corresponding CSLOs were obtained as follows.

- PPI_1: 95th percentile (i.e., 95%) of working time of MRI is 71.0 min.
- $CSLO_1$: Working time of MRI must be less than **71.0** min in **95%** of patients.
- PPI_2: Median (i.e., 50th percentile) of waiting time of MRI is 7223.5 min.
- $CSLO_2$: Waiting time of MRI must be less than **7223.5** min in **50%** of patients.
- PPI_3: Median of waiting time of CT is 4314.5 min.
- $CSLO_3$: Waiting time of CT must be less than **4314.5** min in **50%** of patients.

4.3   Setup for Optimization

Based on the calculated CSLOs, we built a GP optimization model for two activities ($i = \{1 : MRI, 2 : CT\}$) and a couple of time measures ($m = \{wo : working, wa : waiting\}$) in this case study. As the inputs for the GP model, we first used the target values of CSLOs that were derived in Sect. 4.2: $CSLO_1 = 71.0$, $CSLO_2 = 7223.5$, and $CSLO_3 = 4314.5$. Second, we employed three improvement actions ($j=1, 2, 3$): employing more resources (*Action 1*), changing resources into more qualified people (*Action 2*), and employing managers (*Action 3*). As we explained earlier, each action has an effect on decreasing the mean and the standard deviation of time values for entity identifiers (i.e. activities in the case study). Among three actions, the action 1 lowers the average of waiting time for activities, while the action 2 reduces the mean of working time and waiting time. On the other hand, the action 3 decreases the standard deviation of working and average of waiting time. In this model, detailed effects and costs of each action are provided in Table 4.

In the table, costs and effects on working time were assumed, while effects on waiting time were calculated from data. The effects on waiting time in action 1 were inferred from the M/M/c model of the queuing theory. With regard to the action 2 and 3, we calculated the reduction of waiting time according to the change in working time.

Lastly, several assumptions were encoded in the model as manager's decisions and business constraints: expected SLOs, bounds for the number of applications

Table 4. Effects and unit costs of each action for MRI and CT

Action	Cost	Effects on MRI				Effects on CT			
		f^{wo}	h^{wo}	f^{wa}	h^{wa}	f^{wo}	h^{wo}	f^{wa}	h^{wa}
1	1600	–	–	$-1187m$	–	–	–	$-1187m$	–
2	400	-1%	–	$-47.62m/-1m$ of f^{wo}	–	-1%	–	$-62.63m/-1m$ of f^{wo}	–
3	550	–	-10%	$-10m/-1\%$ of h^{wo}		–	-10%	$-10m/-1\%$ of h^{wo}	

for each action, planned implement cost, and importance for each goal. Expected SLOs (i.e., manager's target SLOs) were assumed as follows. These values were applied as constraints in the model with the determined range weight ($W = 0.05$).

– $ESLO_1$: Working time of MRI must be less than **69.0** min in **95%** of patients.
– $ESLO_2$: Waiting time of MRI must be less than **7000.0** min in **50%** of patients.
– $ESLO_3$: Waiting time of CT must be less than **3200.0** min in **50%** of patients.

Also, based on the current status of resources, the number of employing resources ($x_{i,1}$) and changing resources into more qualified people ($x_{i,2}$) for each activity were limited as 1 and 3, respectively. Moreover, we assumed that the planned implement cost was 3000 and the importances for all goals were the same as 0.5.

Based on these inputs, we built a GP model. The complete formulation of each goal and the GP model are presented in Table 5.

Table 5. The GP model for optimization

	Goal 1	Goal 2
O.F.	$DSLO_1 = min\ P_{95}(\mu_1^{wo\,'}, \sigma_1^{wo\,'^2})$ $\mu_1^{wo\,'} = \mu_1^{wo} + \sum_{j=1}^3 x_{1,j} f_{1,j}^{wo}$ $\sigma_1^{wo\,'} = \sigma_1^{wo} + \sum_{j=1}^3 x_{1,j} h_{1,j}^{wo}$	$DSLO_2 = min\ P_{50}(\mu_1^{wa\,'}, \sigma_1^{wa\,'^2})$ $\mu_1^{wa\,'} = \mu_1^{wa} + \sum_{j=1}^3 x_{1,j} f_{1,j}^{wa}$ $\sigma_1^{wa\,'} = \sigma_1^{wa} + \sum_{j=1}^3 x_{1,j} h_{1,j}^{wa}$
	Goal 3	
O.F.	$DSLO_3 = min\ P_{50}(\mu_2^{wa\,'}, \sigma_2^{wa\,'^2})$ $\mu_2^{wa\,'} = \mu_2^{wa} + \sum_{j=1}^3 x_{2,j} f_{2,j}^{wa}$ $\sigma_2^{wa\,'} = \sigma_2^{wa} + \sum_{j=1}^3 x_{2,j} h_{2,j}^{wa}$	
	GP model	
O.F. **subject to**	$max\ Z = 0.5\frac{71.0-DSLO_1}{71.0} + 0.5\frac{7223.5-DSLO_2}{7223.5} + 0.5\frac{4314.5-DSLO_3}{4314.5}$ $69.0 \times (1-0.05) \leq DSLO_1 \leq 71.0$ $7000.0 \times (1-0.05) \leq DSLO_2 \leq 7223.5$ $3200.0 \times (1-0.05) \leq DSLO_3 \leq 4314.5$ $0 \leq \sum_{i=1}^2 \sum_{j=1}^3 x_{i,j} \qquad \sum_{i=1}^2 \sum_{j=1}^3 x_{i,j} = integer$ $\sum_{i=1}^2 x_{i,1} \leq 1 \qquad \sum_{i=1}^2 x_{i,2} \leq 3$ $\sum_{i=1}^2 \sum_{j=1}^3 x_{i,j} c_{i,j} \leq 3000$	

4.4   Optimization Results

Based on the constructed GP model, we obtained the optimal solution. Table 6 provides the optimization results for the case study. The results of optimization with the GP model recommended changing two resources into more qualified people (Action 2) and employing a manager (Action 3) for MRI. Moreover, for CT activity, employing one more resource (Action 1) was suggested. As such, the total used implement cost was turned to 2950. Also, through the optimization, all SLOs were improved. For example, the target value of $CSLO_1$ went from 71.0 min to 68.3 minutes. Likewise, the target values of $CSLO_2$ and $CSLO_3$ were decreased by 141.7 and 1181.1 min, respectively. Lastly, as a result of the combination of importance for each goal, there was a 16.7% reduction.

Table 6. Optimization results for the case study

Applied improvement actions
Changing resources into more qualified people (Action 2) for MRI: 2 (times)
Employing managers (Action 3) for MRI: 1
Employing more resources (Action 1) for CT: 1
Total used cost
2950 (= $400 \times 2 + 550 \times 1 + 1600 \times 1$)
Derived SLOs
$DSLO_1$: Working time of *MRI* must be less than **68.3** min in **95%** of patients
$DSLO_2$: Waiting time of *MRI* must be less than **7076.4** min in **50%** of patients
$DSLO_3$: Waiting time of *CT* must be less than **3133.4** min in **50%** of patients

The result provided the optimal solutions in the given limited cost. In other words, it suggested the best answers for solving the problem that the current process has. Therefore, managers can acquire the direct improvement effects by applying the recommended actions into the activities in the process.

5   Related Work

Numerous research efforts have focused on proposing models for SLA definition in computational and non–computational domains [2,9,10], however, none of them deals with the definition of challenging while achievable SLOs. Some work has been carried out in this direction in the context of computational services. [11] proposes a methodology to calculate SLO thresholds to sign IT services SLAs according to service function cost from a business perspective, but it is useful only for SLAs that apply to the software infrastructure that supports business processes and not for business processes offered as a service. [12] describes a categorization of IT services and outlines a mechanism to obtain efficient SLOs for them. However, they do that at a conceptual level and do not detail how they can be formalized to enable their automated computation.

Regarding the definition of data-based target values or thresholds for PPIs, [13] presents an approach to determine PPI thresholds based on the relationship of different PPIs and their values computed from the process execution data. In this approach, though, a proven relationship between certain PPIs is required in order to extract their thresholds.

Concerning our SLO optimization proposal, some related works exist in the context of process measurement and improvement. A series of proposals exist, e.g. [4,14,15], that identify correlations between PPIs that, eventually, can lead to the definition of process improvement actions. Also related to this is the business process redesign area, which tackles the radical change of a process to enhance its performance dramatically. In this area, a number of works have been presented where heuristic-based BPR frameworks, methodologies, and best practices have been proposed [16,17]. The main drawback of these works concerning our motivating problem is that they are not SLA-aware and leave out of their scope the establishment of target values for the performance measures, or SLOs in the context of business processes offered as services.

6   Conclusion

This paper proposes a structured framework to define realistic SLAs with a systematic evidence-driven approach. The evaluation results obtained from its application to an examination process in the outpatient clinic have shown its applicability and the improvements on the performance of that process.

Our work has a couple of limitations and challenges. The case study adopted for validation covered only the time-related measures. Therefore, a more comprehensive approach that handles various indicators such as frequency and quality is required. Also, with regard to improvement actions in the optimization part, we applied assumptions about the types of actions, costs, and effects. As future work, we will establish more systematic improvement actions by exploring existing works and conducting interviews. In addition, we used the normal distribution-based percentile function with the normality principle. However, if we use the distribution itself (e.g., histogram), we can apply more improvement actions that modify skewness or kurtosis. Therefore, we need to develop a method to support this idea and be able to formulate those improvement actions.

Furthermore, at the beginning, we claimed that our approach aims at reducing the human involvement in the specification of SLOs, but we still need the experts for some steps to gather relevant information. Therefore, we plan to improve our approach by minimizing the human involvement as much as possible and increasing the portion of the data analysis. Finally, in this paper, we focused on the first three steps of the proposed framework. We are already working on implementing the remaining steps and a tool that supports the whole structure. Also, more case studies with real data in different contexts will be performed for further validations.

References

1. Harmon, P.: The scope and evolution of business process management. In: Brocke, J., Rosemann, M. (eds.) Handbook on Business Process Management, vol. 1, pp. 169–194. Springer, Heidelberg (2010). doi:10.1007/978-3-642-00416-2_3

2. del-Río-Ortega, A., Gutiérrez, A.M., Durán, A., Resinas, M., Ruiz–Cortés, A.: Modelling service level agreements for business process outsourcing services. In: Zdravkovic, J., Kirikova, M., Johannesson, P. (eds.) CAiSE 2015. LNCS, vol. 9097, pp. 485–500. Springer, Cham (2015). doi:10.1007/978-3-319-19069-3_30

3. Alves, T.L., Ypma, C., Visser, J.: Deriving metric thresholds from benchmark data. In: 26th IEEE International Conference on Software Maintenance (ICSM 2010), pp. 1–10 (2010)

4. de Leoni, M., van der Aalst, W.M.P., Dees, M.: A general process mining framework for correlating, predicting and clustering dynamic behavior based on event logs. Inf. Syst. **56**, 235–257 (2016)

5. Aouni, B., Kettani, O.: Goal programming model: a glorious history and a promising future. Eur. J. Oper. Res. **133**, 225–231 (2001)

6. Wichura, M.J.: Algorithm as 241: the percentage points of the normal distribution. J. R. Stat. Soc. Ser. C (Appl. Stat.) **37**(3), 477–484 (1988)

7. Whitley, E., Ball, J.: Statistics review 2: samples and populations. Crit. Care **6**(2), 143 (2002)

8. Rozinat, A., Mans, R., Song, M., van der Aalst, W.M.P.: Discovering simulation models. Inf. Syst. **34**(3), 305–327 (2009)

9. Cardoso, J., Barros, A., May, N., Kylau, U.: Towards a unified service description language for the internet of services: requirements and first developments. In: 2010 IEEE International Conference on Services Computing (SCC), pp. 602–609, July 2010

10. Wieder, P., Butler, J., Theilmann, W., Yahyapour, R. (eds.): Service Level Agreements for Cloud Computing, vol. 2506. Springer, New York (2011). doi:10.1007/978-1-4614-1614-2

11. Sauvé, J., Marques, F., Moura, A., Sampaio, M., Jornada, J., Radziuk, E.: SLA design from a business perspective. In: Schönwälder, J., Serrat, J. (eds.) DSOM 2005. LNCS, vol. 3775, pp. 72–83. Springer, Heidelberg (2005). doi:10.1007/11568285_7

12. Kieninger, A., Baltadzhiev, D., Schmitz, B., Satzger, G.: Towards service level engineering for IT services: defining IT services from a line of business perspective. In: 2011 Annual SRII Global Conference, pp. 759–766, March 2011

13. del-Río-Ortega, A., García, F., Resinas, M., Weber, E., Ruiz, F., Ruiz-Cortés, A.: Enriching decision making with data-based thresholds of process-related KPIs. In: Dubois, E., Pohl, K. (eds.) CAiSE 2017. LNCS, vol. 10253, pp. 193–209. Springer, Cham (2017). doi:10.1007/978-3-319-59536-8_13

14. Rodriguez, R.R., Saiz, J.J.A., Bas, A.O.: Quantitative relationships between key performance indicators for supporting decision-making processes. Comput. Ind. **60**(2), 104–113 (2009)

15. Diamantini, C., Genga, L., Potena, D., Storti, E.: Collaborative building of an ontology of key performance indicators. In: Meersman, R., et al. (eds.) OTM 2014. LNCS, vol. 8841, pp. 148–165. Springer, Heidelberg (2014). doi:10.1007/978-3-662-45563-0_9

16. Mansar, S.L., Reijers, H.A.: Best practices in business process redesign: validation of a redesign framework. Comput. Ind. **56**(5), 457–471 (2005)
17. Watson, H.J., Wixom, B.H.: The current state of business intelligence. Computer **40**(9), 96–99 (2007)

A Template for Categorizing Business Processes in Empirical Research

Daniel Lübke[1,2]([⊠]), Ana Ivanchikj[3], and Cesare Pautasso[3]

[1] innoQ Schweiz GmbH, Cham, Switzerland
daniel.luebke@innoq.com
[2] FG Software Engineering, Leibniz Universität Hannover, Hanover, Germany
[3] Software Institute, Faculty of Informatics, USI Lugano, Lugano, Switzerland
ana.ivanchikj@usi.ch

Abstract. Empirical Research is becoming increasingly important for understanding the practical uses and problems with business processes technology in the field. However, no standardization on how to report observations and findings exists. This sometimes leads to research outcomes which report partial or incomplete data and make published results of replicated studies on different data sets hard to compare. In order to help the research community improve reporting on business process models and collections and their characteristics, this paper defines a modular template with the aim of reports' standardization, which could also facilitate the creation of shared business process repositories to foster further empirical research in the future. The template has been positively evaluated by representatives from both BPM research and industry. The survey feedback has been incorporated in the template. We have applied the template to describe a real-world executable WS-BPEL process collection, measured from a static and dynamic perspective.

Keywords: Empirical research · Meta-data template · Business process · Business process description · Business process metrics

1 Introduction

Empirical Research in the field of Business Process Management follows the increasingly wide adoption of Business Process Modeling practices and Business Process Execution technologies [9,17]. The validation of theoretical research, the transfer between academia and industry, and the quest for new research perspectives are all supported by empirical research, e.g., experiments, case studies, and surveys.

The goal of empirical research is to find repeatable results, i.e., observations that can be replicated thus providing results that can be combined and built upon. The more data points are available, the higher the significance of a study. One way to increase the number of data points is to perform meta-studies that combine results from multiple researchers (e.g., [14]). While this is common in

© Springer International Publishing AG 2017
J. Carmona et al. (Eds.): BPM Forum 2017, LNBIP 297, pp. 36–52, 2017.
DOI: 10.1007/978-3-319-65015-9_3

other disciplines, such as ecology or medicine, business process-related data is usually not published in a comparable nor reusable way.

Additionally, the access to industry data is often restricted due to confidentiality requirements. Thus, publication of data sets must be done in an aggregated and/or anonymized manner.

To improve the reporting of empirical research concerning business processes, we propose a template that can be used to characterize processes in terms of their meta-data and (if applicable) their static and dynamic properties, without revealing confidential details. For example, business process models are used for different modeling purposes such as discussion, analysis, simulation, or execution. Processes are modeled using different languages (e.g., BPMN, BPEL, EPC). Process models also vary in terms of their size and structural complexity, which can be determined depending on the actual modeling language used to represent them.

The goal of the proposed template is to (a) give readers the opportunity to "get a feeling" of a process (collection), and (b) allow researchers to build on top of existing research by ensuring the presence of meta-data with well-defined semantics. Since, to the best of our knowledge, no such classification exists, in this paper we make an initial top-down proposal, intended as a starting point for extending and refining the template together with the research community.

In order to improve the reporting of research related to business process model collections (e.g., [6, 20] as a starting point), we propose a set of meta-data described in tabular form. The meta-data template can be extended with other tables. For such extensions we initially propose static metrics for BPEL processes and some dynamic metrics, although further extensions for other modeling languages are welcomed.

We validate the meta-data template by a survey gathering the feedback of academic and industry professionals. Additionally, we apply the template in an industry case study to describe a large process collection.

The remainder of this paper is structured as follows. In Sect. 2 we motivate the need for such template, which we describe in Sect. 3. Section 4 depicts how we validated the template with a survey and a case study. Section 5 presents related work before concluding in Sect. 6.

2 Motivation

Models describing business processes contain sensitive information, making it difficult for companies to reveal how they use standard languages and tools, and rendering it challenging for empirical researchers to further improve the state of the art. As one of our survey respondents emphasized, much of the "research stops at the toy example level."

It is possible to anonymize process models, thereby limiting the understandability of what the process does and hiding their purposes and sources. Anonymized processes retain their entire control and data flow structure (which would be available for static analysis) while loosing important meta-data (which would limit the types of analyses that can be performed).

For example, Hertis & Juric published a large study with a set of over 1'000 BPEL processes [8]. However, they state that they "were unable to classify the processes into application domains since plain BPEL processes do not contain required information." This shows that researchers had to be aware when collecting the processes that they also need to collect associated meta-data.

Thus, whether or not a complete or anonymized process model is present, it is necessary to accompany it with a given set of meta-data. The meta-data has to be carefully selected and placed in a template to ensure that readers and other researchers can get an overall understanding of the discussed processes. Such a template needs to support the following goals:

1. Help researchers to collect data about processes that is relevant to others;
2. Help researchers to publish meaningful results by knowing which properties of the business processes can be anonymized and which should not;
3. Help researchers to report the important properties of business processes in their publications, so that their audience has sufficient details to evaluate the quality of the reported research;
4. Foster empirical research about business processes so that a body of knowledge can be accumulated based upon multiple, comparable works;
5. Enable meta-studies that combine, aggregate and detect trends over existing and future empirical research about practical use of business processes.

3 Template

Business Process Models can be created in many languages and can serve many purposes. Thus, it makes sense to report only values that have been actually measured in the specific usage context and are related to the conducted research. The templates are defined in a tabular format with a key/value presentation in order to allow quick digestion and comparison of reports. We understand that research publications need to present their results in a compact form. When space does not allow to use the tabular format, the tabular templates can be published together with the data, e.g., in technical reports and research data repositories.

The template we propose is built in a modular fashion. It consists of a required meta-data template that describes general, technology-independent properties of the process. The meta-data part can be extended by standardized templates for reporting different properties that have been analyzed. Researchers should re-use existing templates as much as possible in order to provide results that can be compared to previous works.

For instance, in this paper, two additional templates for executable BPEL processes are presented. The list of static and dynamic metrics proposed in the additional templates is not exhaustive and can be extended depending on the research needs. BPEL was chosen for convenience, as the case study in Sect. 4.2 uses BPEL processes. Support for other languages can be easily defined in additional templates.

3.1 Meta-data Template

The meta-data template, as shown in Table 1, is the only required part. It is designed to be applicable to any process model regardless of the modeling language used. This template contains the basic information necessary to obtain general understanding about a process model and the most important properties that can be of interest to filter and classify such process model. Its content has been updated with the feedback received during the survey described in Sect. 4.1. Following is a more detailed description of the categories and the classes included in the table:

Table 1. The meta-data template for describing business process models

Process Name	Name or Anonymous Identifier of the Process
Version	Process Version (if available)
Domain	Business Domain of the Process
Geography	Location of the processes
Time	Period of data collection
Boundaries	Cross-Organizational/Intra-Organizational/Within-Department
Relationship	Calls another/Is being called/No call/Event triggered
Scope	Business Scope: Core/Auxiliary, or Technical Scope
Process Model Purpose	Descriptive/Simulation/Execution
People Involvement	None/Partly/No Automation
Process Language	e.g., WS-BPEL 2.0/EPC/BPMN1/BPMN2/...
Execution Engine	Engine used for running the Process Model if the model is executable
Model Maturity	Illustrative/Reference/Prototypical/Reviewed/Productive/Retired

Process Name: The process name as used in the organization. If the real name cannot be published, this field can be anonymized by providing an ID that can be used to reference the process from the text;

Version: If available, the name can be augmented with process versioning meta-data;

Domain: The business domain which this process is taken from. Existing ontologies like [7] can be used;

Geography: The geographical location where the process is used;

Time: The time period the process data refers to;

Boundaries: The organizational scope of the process: *cross-organizational* for processes that span across multiple legal entities, *intra-organizational* for processes that are conducted within one legal entity but across different departments/units in it and *within-department* for processes that are narrowed to a single organizational unit within one legal entity;

Relationship: The structural dependencies of the process with other processes: *calls another, is being called, no call, event triggered*;

Scope: The process model can have a horizontal, business scope, or a technical scope. In the business scope we can distinguish between: *End-to-end* processes for fully end-to-end descriptions like order-to-cash, and *auxiliary* processes for processes that do not contribute directly to the business purpose. Processes can have a pure technical scope instead, e.g., an event handling process that propagates permissions in the infrastructure;

Process Model Purpose: The purpose of a process model can be description, simulation or execution. A *descriptive* process is a model from a business point of view, which is more abstract in order to facilitate discussion and analysis among stakeholders, and also to prescribe how operations are carried out in an organization; a *simulative* process contains further details regarding resources, costs, duration, frequency, etc., while an *executable* process contains sufficient details to enable the automation of the process. Because a model can serve multiple purposes, this field is a list. The main purpose should be the first item in this list;

People Involvement: Classification of how much manual/human work is to be done. Ranges from *none* (fully automated) over *partly* to *no automation* (people involvement in each task);

Process Language: The process language used to create the process model. If a standard process language, such as BPEL, BPMN, etc., has been extended that should be specified in the meta-data;

Execution Engine: The execution engine(s) used to run the process model (if executable), including the exact version, if available;

Model Maturity: *Illustrative* for models which are not intended for industry use but to showcase certain modelling situations for educational purposes, *reference* for generic models which prescribe best practices and are used as starting point for creating other types of models, *prototypical* for models that are under discussion or are technical prototypes, *reviewed* for models that have been reviewed but are not yet in productive use, *productive* for models that are used productively in a real-world organization, with or without systems to enact them automatically, and *retired* for models which had been productive previously but have been replaced with other models.

The meta-data template is the main template that describes process characteristics regardless of the context and used technologies. In order to report details, additional templates should be used which often need to be language specific. Within this paper we define additional templates that describe different viewpoints of business processes, especially for those modeled in executable WS-BPEL.

3.2   BPEL Element and Activity Count Template

One of the interesting properties of processes are the various "size" metrics, with "size" being defined by Mendling [13] as "often related to the number of nodes N of the process model." Since every process language provides different ways to express nodes and arcs for defining the control-flow, such template must be process language-specific. Thus, in this paper we define the template for measuring the size of BPEL processes by using activity and element counts, since BPEL is used in the case study that is presented in Sect. 4.2.

The template for reporting BPEL Element Counts is shown in the case study in Table. 3. The values are merely the counts of different BPEL constructs as defined by the WS-BPEL 2.0 standard [10]. In addition, the total count of basic activities and structured activities is given because these are often used to judge the size of a process model. In the literature they are also called Number of Activities (NOA) and Number of Activities Complex (NOAC) [5]. In addition to activities, this table also contains the number of links, number of different sub-activity constructs (e.g., pick branches, if branches), and the number of partner links (service partners). To distinguish between the different BPEL constructs, basic activities are marked with a (B) and structured activities are marked with an (S) in Table 3.

3.3   BPEL Extensions Template

Although BPEL is a standardized language, it offers support for extensions. These extension points are used to extend the BPEL standard, e.g., the standardized extension BPEL4People to support human tasks, or to enable vendors to offer unique features that distinguish their products from their competitors'. BPEL defines a general facility to register extensions globally and the extension activity that can contain activities that are not defined in the core standard, or to use additional query and expression languages that are referenced by a non-standard URI. In contrast to [15] we think that the use of extensions is common. Also the case study has shown a high use of both vendor-specific and standardized extensions.

When reporting on BPEL processes, researchers can use the template as shown in the case study in Table 4 that contains all declared extensions in the BPEL process and the extension activities used together with their activity counts.

3.4   Process Runtime Performance Template

For executable processes, it becomes possible to report their runtime performance. While a large number of metrics have been proposed (e.g., [18]), for space reasons, in this paper we propose to focus on reporting the number of process instances and their duration. These metrics can be described for each process instance or aggregated among multiple instances.

Counting the total number of process instances for a given process model gives an idea of its usage frequency relative to other process models.

Capturing the performance of individual process instances amounts to measuring their execution time $(T(finish) - T(start))$. Since the execution time of every process instance is usually not of interest, we suggest to give statistical information about the distribution of the process instance duration for all process instances of a given process model as shown in Table 5.

4   Validation

To validate the usefulness of the proposed templates we combine an exploratory survey with researchers and industry experts (Sect. 4.1) and a case study of real-world BPEL business processes (Sect. 4.2).

4.1   Survey with Researchers and Industry Experts

To validate whether the proposed template fulfills the goals presented in Sect. 2 we have conducted an exploratory survey [19, Chap. 2][1]. The intention of this survey was not statistical inference of the results, but rather getting a deeper understanding of the surveyed field. We targeted audience from both academia and industry, i.e., both producers and consumers of empirical research. Thus, we used different social media channels and private connections to disseminate the survey.

Survey Design. We organized the questions in five sections: Background, Meta-Data Template, Template Remarks, Template Extensions and Empirical Research in BPM. While the Background questions were mandatory to enable further classification in the analysis of the results, the remaining questions were optional to incentivize greater survey participation. In the Meta-Data Template section we showed the meta-data presented in Table 1 and asked the respondents to rate the importance of each of the proposed meta-data classes. In the Template Remarks section we focused on the perceived need of standardized reporting and asked suggestions for the appropriateness and completeness of the proposed process classification and meta-data. In the Template Extensions section we inquired about the relevance of reporting structure and performance metrics on process level, as well as on the usefulness of using the meta-data and metrics for describing entire collections of process models. Last but not least, in the Empirical Research in BPM section we asked for personal opinions on the state of the empirical BPM research.

Survey Sample. Since we were not aiming at inferring statistical conclusions from the conducted survey, we closed the survey as soon as we considered the obtained feedback sufficient for improving the proposed templates. This has resulted with 24 respondents with diversified background. To obtain more insights into respondents' professional background, they could select multiple options between experience in academia (further divided into IT or Business Process Management), and in industry (further divided into IT or Business). While most of the respondents, i.e., 46% have experience only in academia, 21% have experience only in industry and 33% in both academia and industry. Most of them, i.e., 88% have IT background (16 respondents in academia and 12 in industry) and 63% have been dealing with the business perspective of process management (12 respondents in academia and 3 in industry).

---

[1] The questionnaire is available at http://benchflow.inf.usi.ch/bpm-forum-2017/.

Respondents participate in different phases of the business process life-cycle, and/or simply conduct empirical research on BPM. When asked what type of experience they have with business processes, the majority, i.e., 83% marked analyzing, while 79% marked defining, 75% implementing and 29% researching. These results could already indicate some lack of empirical research in this area.

All the respondents have more than one year of experience in working with business processes with 50% having up to 5 years and other 33% over 10 years of experience. Figure 1 shows the years of experience vs. the business process life-cycle experience of the survey participants. It is noticeable that people with longer experience have been more exposed to different phases of the business process life-cycle.

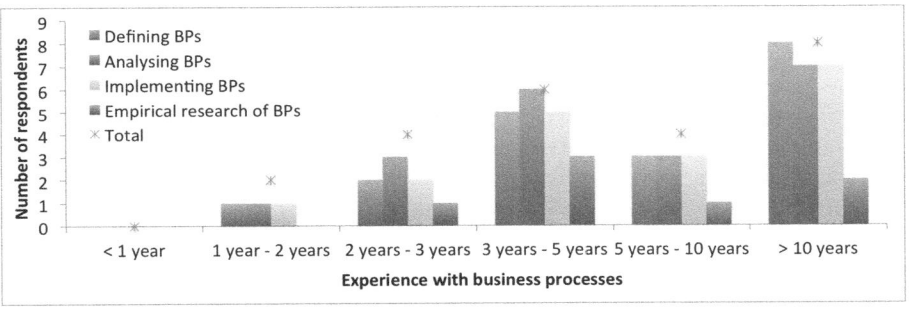

Fig. 1. Survey respondents: years of experience vs. business process areas expertise

Survey Results. We have presented the meta-data and process classifications as shown in Sect. 3.1 to the respondents, which in addition included the Modeling Tool category that we removed from the updated table as per respondents' feedback. We asked them to evaluate each proposed category on a scale from 1 (not important) to 5 (very important). As per the average score the Process Model Purpose is considered the most important with 4.38 points to be followed by People Involvement with 4.13 points. As mentioned previously, the Modeling Tool was considered as the least valuable with 3.17 points together with the Execution Engine with 3.38 points. Indeed in an ideal world, where the standards are correctly implemented, these two categories would not add to the understanding of the process model. In Fig. 2 we stratify the importance rating of each proposed category per sector (industry, academia or both). It is interesting to notice that, even if those having experience only in industry allocate less importance to the meta-data on average, similar importance trends are evident between the different sectors. If stratified per years of experience, the highest ratings are provided by respondents with 1 to 2 years of experience to be followed by those with over 10 years of experience.

Encouraging ratings were also obtained on the helpfulness of the standardized reporting approach for "getting a feeling" about the studied process (4.08

points on average) and for comparing different empirical reports (4.29 points on average). Based on the feedback on missing meta-data we have added the Version, Geography, Time, and Relationship categories to Table 1 as well as the Reference and Retired classes in the Model Maturity category.

In the next section of the survey we focused on the extended tables presented in Sects. 3.2 and 3.4. Always on the same scale from 1 to 5, the respondents found the presentation of structure metrics and performance metrics sufficiently relevant, with average points of 3.40 and 3.57 respectively. We were curious to see whether priorities and interests change when using the meta-data and extended data presented in Sect. 3 on a collection of business processes. Thus, we asked respondents to rate them. While on process level, as mentioned earlier, Process Model Purpose and People Involvement were considered the most important, at collection level the Aggregated Structured Metrics (4.11 points) and the Domain (3.84 points) were considered the most important. As on process level, on collection level as well, the least important remained the Modeling Tool (3.11 points) and the Execution Engine (2.68 points).

As for the processes, also with the collections the responses followed similar trends among different sectoral experiences (academia, industry or both) evident from Fig. 3, with industry always providing lower average scores than academia, while people with experience in both sectors tending to have opinions more aligned with academia. The greatest differences in opinions between industry and academia refer to the Model Maturity and Process Name where average academia's importance rating is around 4 while industry's importance rating is around 3 on process level and 2 on collection level. Significant differences in opinion are also noticed on collection level regarding the importance of the Structure Metrics which are rated at 2.5 by industry, 3.9 by academia and 4.9 by respondents with experience in both sectors. However, when aggregating among the importance rating of all proposed meta-data and extended data categories, the opinions are relatively positive with an average of 3.77 out of 5 points for data on process level and 3.53 out of 5 points for data on collection level.

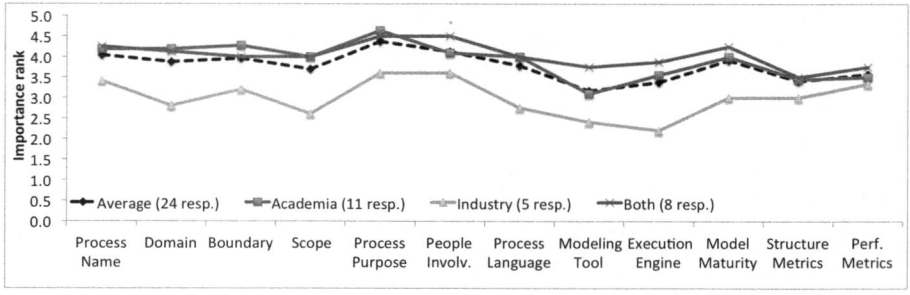

Fig. 2. Process Meta-Data Template validation (mean importance)

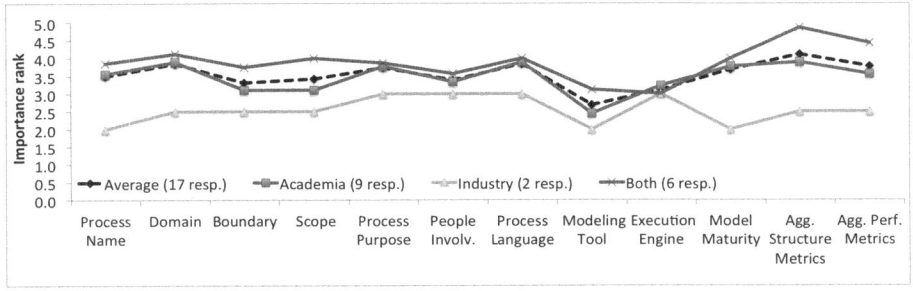

Fig. 3. Process Collection Meta-Data Template validation (mean importance)

We asked for additional properties that respondents would like to have in the template. Two recommendations, the connectedness of the model and a link to a process map, were made. However, connectedness is hard to define without requiring a special modeling language, while without standardized process maps, we think that the links are not helpful.

Last but not least, when asked whether they consider the existing empirical research in business process management (surveys, experiments, case studies) sufficient, out of the 16 respondents to this question only 4 answered positively.

4.2   Case Study with Industry Processes

We use the Terravis project as a case study for reporting process meta-data and metrics in a standardized fashion. Terravis [2] is a large-scale process integration project in Switzerland that coordinates between land registries, banks, notaries and other parties business processes concerning mortgages. In contrast to previous reportings of metrics [11], in this paper we apply our template and all additional templates as defined in this paper.

The Research Questions addressed by this case study are the following:

- Can the template be applied without problems? Especially are all category values clearly defined and applicable?
- Can all categories be measured? Which measurements can be automated?
- Is the categorization in the meta-data template beneficial when evaluating the process metrics?

The set contains 62 executable BPEL models that are executed on ActiveVOS 9.2. We could acquire a total of 918 versions of the process models and information for 435,093 process instances executed in Switzerland in the period between 2012 and 2016. To apply the templates we conducted the following steps:

1. For each process we assigned a value to each category of the general meta-data template, automating the assignment where possible;
2. Automatically measured the static metrics for the models;

3. Validated the People Involvement assignment by cross-checking the value of the count of human activities in the static metrics;
4. Automatically collected the used BPEL extensions; and
5. Calculated the run-time metrics from the process logs.

In the first step we manually classified each process as per our meta-data template. In the People Involvement category we initially chose to offer more fine-grained values (partly, mostly). However, it was impossible to find a meaningful and objective threshold for these values. Thus, we opted to offer only one intermediate value, i.e., partly. To show-case the application of the meta-data template the meta-data of one process model is shown in Table 2.

Table 2. The meta-data template for a terravis process

Process Name	Transfer Register Mortgage Certificate to Trustee
Version	26.0
Domain	Land Registry Transactions
Geography	Switzerland
Time	2016-08-30
Boundaries	Cross-Organizational
Relationship	Calls another/Is being called
Scope	Core
Process Model Purpose	Executable
People Involvement	None
Process Language	WS-BPEL 2.0 plus vendor-extensions
Execution Engine	Informatica ActiveVOS 9.2
Model Maturity	Productive

Many static metrics, e.g., the static element counts [3,12] have been proposed and some tools have been developed for calculating them [1,8]. However, to our knowledge no working tool is freely available to calculate element counts and extract extension information from BPEL process models. Thus, we have built an open source implementation[2] to automatically calculate the data for the BPEL element and activity count template (Table 3).

To calculate the run-time metrics, the process logs were extracted and processed automatically. However, not all executable processes were configured with persistence and logging enabled. Thus, for some models we could not calculate any run-time metrics. Process instance run-time metrics are shown in Table 5.

After successfully applying the templates to all process models, an aggregation over the whole collection can be made. The results are shown in templated

---

[2] Available at https://github.com/dluebke/bpelstats.

Table 3. BPEL element and activity counts for a terravis process

Transfer register mortgage certificate to trustee (Version 26.0)			
BPEL element	Count	BPEL Element	Count
Assign (B)	79	OnAlarm (Pick)	0
Catch	4	OnAlarm (Handler)	0
CatchAll	2	OnMessage (Pick)	6
Compensate (B)	0	OnEvent (Handler)	0
Compensate Scope	0	Partner Link	15
Compensation Handler	0	Pick (S)	3
Else	13	Receive (B)	13
Else If	3	Repeat Until (S)	0
Empty (B)	42	Reply (B)	18
Exit (B)	9	Rethrow (B)	0
Extension Activity	1	Scope	74
Flow (S)	1	Sequence (S)	90
ForEach (S)	4	Throw (B)	0
If (S)	13	Validate (B)	0
Invoke (B)	37	Wait (B)	0
Link	2		
Derived Metrics:			
Basic Activities (B)	198	Structured Activities (S)	185

form in Table 6 with information on the percentage of models belonging to each class.

If the categorization in the meta-data template is meaningful, there should be no overlapping between classes in the same category and preferably each class should have some processes which pertain to it. We grouped the static metrics and process duration metrics of the latest version of every process model according to the different categories and their classes. The results are shown in Table 7. As can be seen, the distribution of the number of process models in the classes is different than the distribution of the number of activities. For example, only 37% of the process models describe cross-organizational processes but they contain 71% of the activities. This means that on average the cross-organizational models are larger than those in the different classes of the Boundaries category, and the within-system processes are the smallest on average. The distribution of the number of process instances and the distribution of the accumulated process duration among all executed process instances also differ. Only 14% of the process instances are cross-organizational but account for 68% of the overall process time spent. This means that cross-organizational and intra-organizational processes on average take longer to complete than within-system processes. Also technical process models have a very different distribution.

Table 4. BPEL extensions for a terravis process

Extensions	http://www.activebpel.org/2006/09/bpel/extension/activity
	http://www.activebpel.org/2009/06/bpel/extension/links
	http://www.activebpel.org/2006/09/bpel/extension/query_handling
	http://www.activebpel.org/2009/02/bpel/extension/ignorable
	http://www.omg.org/spec/BPMN/20100524/DI
Activities	Type	Count
	ActiveVOS Continue	1
	Total	1

Table 5. Template for capturing run-time performance metrics of process instances

Transfer Register Mortgage Certificate to Trustee (Version 26.0)	
Number of Process Instances	13
Execution Time (min)	00h:00m:01s
Execution Time (med)	02h:33m:00s
Execution Time (mean)	12h:34m:39s
Execution Time (max)	64h:24m:14s
Execution Time (total)	163h:30m:32s

Table 6. Aggregated meta-data for the terravis process collection

Collection Name	Terravis
Process Count	62 Models with 918 versions
Domain	Land Registry Transactions
Geography	Switzerland
Time	2012-03-09 – 2016-08-30
Boundaries	Cross-Organizational 37%, Intra-Organizational 13%, Within-System 50%
Relationship	Is being called 31%, Calls another 26%
	Is being called/Calls another 8%, Event triggered 24%
	No call 11%
Scope	Technical 52%, Core 39%, Auxiliary 10%
Process Model Purpose	Executable
People Involvement	None 79%, Partly 21%
Process Language	WS-BPEL 2.0 plus vendor-extensions
Execution Engine	Informatica ActiveVOS 9.2
Model Maturity	51 Productive, 11 Retired Models
	51 Productive, 867 Retired Model Versions

Table 7. Distribution of terravis process models and instances by category

	#Model	#Activities	#Instances[a]	#Duration
Total	62	10'132	86'035	2'238'583 h
Boundary				
Cross-Organizational	37%	71%	14%	68%
Intra-Organizational	13%	19%	8%	32%
Within-System	50%	10%	78%	0.1%
Relationship				
Is being called	31%	22%	19%	71%
Calls another	26%	55%	62%	9%
Is being called, Calls another	8%	12%	2%	20%
Event triggered	24%	3%	15%	0%
No call	11%	9%	2%	1%
Scope				
Technical	52%	10%	85%	0.2%
Core	39%	85%	13%	99%
Auxiliary	10%	5%	2%	1%
People Involvement				
None	79%	66%	86%	10%
Partly	21%	34%	14%	90%
Model Maturity				
Production	82%	84%	100%	96%
Retired	18%	16%	0.2%	4%

[a] Only for latest process model version.

The results support the classification categories because based on these values different characteristics of the processes in this collection are exhibited.

5   Related Work

The extensions to the meta-data template (Sects. 3.2, 3.3 and 3.4) are language specific, and their aim is emphasizing the need of including structure and performance metrics, while not trying to be exhaustive in the list of metrics. Defining such metrics is out of the scope of this paper, and has already been addressed in existing work [4,5,13,18]. The main goal of this paper is standardizing the metadata on process model and/or collection level. Thus, the related work we survey in this section refers to current availability and definition of such meta-data.

The need of extracting knowledge from business processes has been identified in literature and has led to the creation of business process repositories. Yan et al. [20] propose a Repository Management Model as a list of functionalities that can be provided by such repositories and survey which of them are offered by existing repositories. Since what they propose is a framework, they emphasize the need of meta-data for indexing the processes, but do not define which meta-data should accompany each process. They have found that only 5 out of 16 repositories use a classification scheme based on part-whole and generalization-specialization relations. Vanhatalo et al. [16] built a repository for storing BPEL processes with the related meta-data, which in their usage scenario referred to the: number of activities, degree of concurrency, execution duration and correctness. Their flexible repository architecture could be used to store the templates proposed in our paper. The MIT Process Handbook project focuses on classifying the process activities and on knowledge sharing[3]. We focus on standardization of the reporting of such acquired knowledge.

The BPM Academic Initiative [6] is a popular process repository offering an open process analysis platform, aimed at fostering empirical research on multiple process collections. The meta-data required when importing processes refers to the process title, the collection it belongs to, the process file format and modelling language. Even though the data to be stored is not restricted only to these fields, no further standardization of the process classification is offered. In their survey on empirical research in BPM, Houy et al. [9] define a meta-perspective, a content-based and a methodological perspective for classifying the surveyed articles. Their content-based perspective refers to context (industry or public) and orientation (technological, organizational or inter-organizational). The standard meta-data we propose can offer a richer classification for meta-studies like [9, 14] and more in-depth analysis performed using platforms like [6].

6   Conclusions and Future Work

Empirical research in BPM helps to close the feedback loop between theory and practice, enabling the shift from assumptions to facts and fostering real-world evaluation of so far untested theories. While the process mining research has benefited from the availability of large event log collections, the same cannot be claimed concerning process model collections [6]. As process models clearly represent trade secrets for the companies using them productively, in this paper we have proposed a language-independent template for describing them by focusing on key properties (classification meta-data, size & instance duration) which are useful for empirical analysis by the academic research community without revealing proprietary information. The template has been validated with an exploratory survey among 24 experts from industry and academia, who have positively commented on the choice of properties (no negative score was reported) and also made constructive suggestions that have already been incorporated in the template described in this paper. We have also demonstrated the applicability of the template in an industrial case study by using it to report on the Terravis collection of 62 BPEL processes and a subset of their 435,093 process instances executed across multiple Swiss financial and governmental institutions in the period between 2012 and 2016.

---

[3] http://process.mit.edu/Info/Contents.asp.

While the meta-data template presented in this paper is language independent, the extensions concerning static metrics are BPEL specific. Therefore, we plan to work on similar templates for other modeling languages in the future. Additionally, we plan to collaborate with modeling tool vendors to enable the automated collection of the meta-data described in this paper. The long-term plan is to grow the amount of available and well-classified process models to the empirical BPM community. One way to increase the number of classified processes is to auto-classify existing model collections. Future work will elaborate which properties can be inferred from existing data.

Most of the respondents of our survey said that there is not enough empirical research in the field of BPM. We hope that more empirical research will be conducted and that the meta-data presented in this paper will help researchers to improve the classifications of data collections and make them easier to compare and re-use across different publications.

Acknowledgments. The authors would like to thank all of the participants in the survey for their time and valuable feedback.

References

1. Alemneh, E., et al.: A static analysis tool for BPEL source codes. Int. J. Comput. Sci. Mob. Comput. **3**(2), 659–665 (2014)
2. Berli, W., Lübke, D., Möckli, W.: Terravis - large scale business process integration between public and private partners. In: Plödereder, E., Grunske, L., Schneider, E., Ull, D. (eds.) Lecture Notes in Informatics (LNI), vol. P-232, pp. 1075–1090. Gesellschaft für Informatik e.V. (2014)
3. Cardoso, J.: Complexity analysis of BPEL web processes. Softw. Process Improv. Pract. J. **12**, 35–49 (2006)
4. Cardoso, J.: Business process control-flow complexity: Metric, evaluation, and validation. Int. J. Web Serv. Res. (IJWSR) **5**(2), 49–76 (2008)
5. Cardoso, J., Mendling, J., Neumann, G., Reijers, H.A.: A discourse on complexity of process models. In: Eder, J., Dustdar, S. (eds.) BPM 2006. LNCS, vol. 4103, pp. 117–128. Springer, Heidelberg (2006). doi:10.1007/11837862_13
6. Eid-Sabbagh, R.-H., Kunze, M., Meyer, A., Weske, M.: A platform for research on process model collections. In: Mendling, J., Weidlich, M. (eds.) BPMN 2012. LNBIP, vol. 125, pp. 8–22. Springer, Heidelberg (2012). doi:10.1007/978-3-642-33155-8_2
7. Executive Office of the President - Office of Management, Budget: North American Industry Classification System (2017). http://census.gov/naics
8. Hertis, M., Juric, M.B.: An empirical analysis of business process execution language usage. IEEE Trans. Softw. Eng. **40**(08), 738–757 (2014)
9. Houy, C., Fettke, P., Loos, P.: Empirical research in business process management-analysis of an emerging field of research. Bus. Process Manag. J. **16**(4), 619–661 (2010)
10. Jordan, D., et al.: Web Services Business Process Execution Language Version 2.0. OASIS, April 2007
11. Lübke, D.: Using metric time lines for identifying architecture shortcomings in process execution architectures. In: 2015 IEEE/ACM 2nd International Workshop on Software Architecture and Metrics (SAM), pp. 55–58. IEEE (2015)

12. Mao, C.: Control and data complexity metrics for web service compositions. In: Proceedings of the 10th International Conference on Quality Software 2010 (2010)
13. Mendling, J.: Metrics for Process Models: Empirical Foundations of Verification, Error Prediction, and Guidelines for Correctness, 1 edn. Springer (2008)
14. Mendling, J.: Empirical studies in process model verification. In: Jensen, K., Aalst, W.M.P. (eds.) Transactions on Petri Nets and Other Models of Concurrency II. LNCS, vol. 5460, pp. 208–224. Springer, Heidelberg (2009). doi:10.1007/978-3-642-00899-3_12
15. Skouradaki, M., Roller, D., Pautasso, C., Leymann, F.: "bpelanon": Anonymizing bpel processes. In: ZEUS, pp. 1–7. Citeseer (2014)
16. Vanhatalo, J., Koehler, J., Leymann, F.: Repository for business processes and arbitrary associated metadata. In: Proceedings of the Demo Session of the 4th International Conference on Business Process Management (2006)
17. Weber, B., Mutschler, B., Reichert, M.: Investigating the effort of using business process management technology: results from a controlled experiment. Sci. Comput. Program. 75(5), 292–310 (2010)
18. Wetzstein, B., Strauch, S., Leymann, F.: Measuring performance metrics of WS-BPEL service compositions. In: Proceedings of ICNS, pp. 49–56 (2009)
19. Wohlin, C., Runeson, P., Höst, M., Ohlsson, M.C., Regnell, B., Wesslén, A.: Experimentation in Software Engineering. Springer, Heidelberg (2012)
20. Yan, Z., Dijkman, R., Grefen, P.: Business process model repositories-framework and survey. Inf. Softw. Technol. 54(4), 380–395 (2012)

Mining and compliance

Toward a New Generation of Log Pre-processing Methods for Process Mining

Paolo Ceravolo[1], Ernesto Damiani[2], Mohammadsadegh Torabi[1],
and Sylvio Barbon Jr.[3(✉)]

[1] Università degli Studi di Milano, via Bramante 65, Crema, Italy
paolo.ceravolo@unimi.it, mohammadsadegh.torabi@studenti.unimi.it
[2] Khalifa University, PO Box 127788, Abu Dhabi, UAE
ernesto.damiani@kustar.ac.ae
[3] Londrina State University, Rod. Celso Garcia Cid, 445, Londrina, Brazil
barbon@uel.br

Abstract. Real-life processes are typically less structured and more complex than expected by stakeholders. For this reason, process discovery techniques often deliver models less understandable and useful than expected. In order to address this issue, we propose a method based on statistical inference for pre-processing event logs. We measure the distance between different segments of the event log, computing the probability distribution of observing activities in specific positions. Because segments are generated based on time-domain, business rules or business management system properties, we get a characterisation of these segments in terms of both business and process aspects. We demonstrate the applicability of this approach by developing a case study with real-life event logs and showing that our method is offering interesting properties in term of computational complexity.

Keywords: Process mining · Event-log clustering · Pre-processing · Lightweight trace profiling

1 Introduction

The well-known idiomatic expression "garbage in, garbage out" applies well to Process Mining (PM), because significant results can be achieved only if the *event logs* fed into PM algorithms are good examples of execution for all the relevant variants in a business process [21].

This problem is already recognised by the literature and many contributions underline that before running process discovery it is required to pre-process event logs [3,4]. Clustering is considered one of the most relevant pre-processing tasks as grouping similar event logs can radically reduce the complexity of the discovered models [1,5,11,17,18][1]. Despite this attention, the methods proposed

---

[1] Some works, such as for instance [22], define as "Clustering" the identification of similar activities, this is also a pre-processing task relevant to our discussion, however in this paper we are using "Clustering" for referring uniquely to the process of segmenting event logs.

© Springer International Publishing AG 2017
J. Carmona et al. (Eds.): BPM Forum 2017, LNBIP 297, pp. 55–70, 2017.
DOI: 10.1007/978-3-319-65015-9_4

in the literature are only partially tailored to the specific needs of Business Process Management (BPM) [12], where business goals and rules [7] are tailored on each specific business process and monitoring or discovery should be tailored accordingly [10].

Few existing process mining techniques are equipped with means for uncovering differences among event logs. Moreover, with the notable exception of [11], little attention has been devoted to the development of a comprehensive method for coping with the entire work-flow that guides pre-processing tasks. This workflow includes at least the following steps: (i) *characterisation of events logs*, (ii) *computation of a similarity measure* and finally (iii) *evaluation on the business relevance* of the divergences or convergences of the characteristics considered. These tasks cannot be considered in isolation and multiple iterations over them may be required to get significant results.

Differently from the currently adopted log pre-processing practises, the approach we propose in this paper introduces the notion of segment that is a sub set of the event log that conforms to some specific business goal or business rule. Statistical inference-based analysis allows to characterise the distribution of activities in segments, providing an explanation of their similarities or dissimilarities. More specifically the paper is organized as follows: we start introducing the related works in Sect. 2. We then present an overview on our method in Sect. 3[2]. In particular, in Sect. 3.1 we provide preliminary definitions and explain how event logs can be segmented; in Sect. 3.2 we introduce a new trace profiling method that can be exploited in comparing and clustering traces; in Sect. 3.3 we illustrate distance metrics based on inferential statistics; in Sect. 3.4 we discuss how to use our results to characterise segments and evaluate if they are suitable to input process discovery. In Sect. 4 we demonstrate the applicability of our method using a case study with real-life event log from an Italian manufacturing company[3]. In Sect. 5 we compare our method to the state of the art via a time complexity analysis and finally in Sect. 6 we draw our conclusions.

2 Related Work

Just like Data Transformation [2] and Data Cleansing [24], Trace Clustering [19] is a crucial step in pre-processing event logs, as it can radically reduce the inherent complexity of discovered models. Song et al., in [20], present an introduction to trace clustering algorithms with trace profiling. A profile is a set of related items that describe a trace from a specific perspective. These perspectives usually rely on derived information, such as the number of events in a trace or the resources consumed during execution. A profile with n items is a function, which assigns to a trace a vector with n elements. Encoding a

---

[2] The Python implementation of the algorithms adopted to implement and test our method is available at http://www.uel.br/grupo-pesquisa/remid/wp-content/uploads/LightPMClustering.rar.

[3] The event log is available at http://www.uel.br/grupo-pesquisa/remid/wp-content/uploads/EventLogDatasetAnon.csv.

trace into a vector space model makes possible to compute distance metrics and perform cluster analysis.

In [5], authors use a trace clustering approach based on edit distance[4], where profiles are obtained by listing the activities into a trace (bag-of-activities). This is a straightforward approach that offers linear computational complexity when computing a distance measure between traces, but loses all information on the trace structure.

To incorporate information about trace structure, it is possible to adopt contextual approaches. These approaches generate vectors using k-grams [5], i.e. representing each activity in the trace as a sub-sequence of length k. Even though it has been shown that techniques that take into account context perform better than those that do not, the high complexity of k-gram, $\mathcal{O}(n^k)$, is an obstacle respecting most of the state of art methods with linear complexity.

Recent research focuses on extracting multiple trace profiles to exploit multi-criteria clustering techniques. In [11] the authors proposed a framework to deal with the more general correlation problem by a tool that merges previous approaches in the literature. Appice and Malerba, in [1], proposed co-training clustering as a pre-processing step. The output is a trace clustering pattern, obtained by clustering the traces across multiple profiles inputted. The co-training idea is based on iterative modification of a similarity matrix extracted from the trace profile. The time complexity of such an algorithm depends on the cost of computing the similarity and clustering matrices which are respectively $\mathcal{O}(n^2 M)$ and $\mathcal{O}(d)$, where n is the number of traces, M the average number of features per trace profile and d the cost of the distance-based algorithm used.

When computing the similarity between two activities most methods do not deal with semantics, e.g. cannot capture tasks that are expressed using different abstraction levels but refer to the same business activity. Chen et al. proposed in [9] a method which can address semantic aspects as well as structural features of the event log. Their pre-processing method is based on k-means clustering, whose cost is $\mathcal{O}(n^{dk+1})$ over vectors that encode both structure and semantic features, where n is the number of traces, d the cost of distance function and k the number of compared traces. Structure information is derived from control-flow representations such as loops, branches and sequences. It is however important to note that the proposed solution can extract control-flow and semantics features only if a deep pre-processing analysis is performed, thus the challenges we outlined in the introduction are moved rather than solved.

3   Overview on the Proposed Method

An high-level overview of the method proposed in this paper is shown in Fig. 1. Our starting point is an event log that collects a set of cases, i.e. instances of business process execution. To select sets of cases that can meaningfully be fed to process discovery we propose a method organised in four steps. A criteria for

---

[4] The edit distance between two strings is the minimum number of operations required to transform one string into the other.

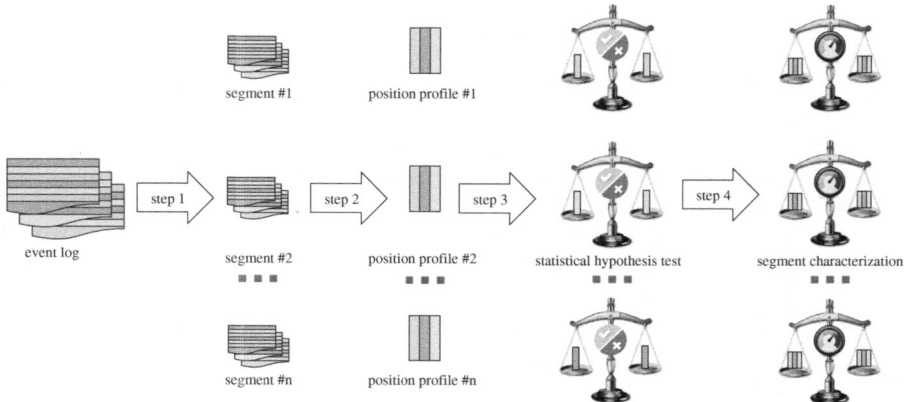

Fig. 1. Overview of method proposed in this research

segmenting the event log is first identified; segments are then *represented based on a trace profile*, where a trace is a unique sequence of events generated in executing a business process, which is adopted to *compute a similarity measure*; finally the adopted segmentation is *assessed and characterised*. In the following we describe each step providing a formalisation of the operations implemented and related examples to clarify the details.

3.1   Step 1: Segmenting the Event Log

The first step is splitting the event log into group of cases called segments. We rely on the event log description standard proposed by the IEEE Task Force on Process Mining. The eXtensible Event Stream (XES) [14] defines a grammar for a language capturing information systems' behaviors. In this framework, an event stream describes a set of events that can be ordered in a sequence using their execution timestamps. More specifically an event can be defined according to Definition 1.

Definition 1. *Event. An event is a quadruple $e = (c, a, r, t) \in \mathcal{E}$, denoting the occurrence of an activity a in a case c, using the resource r at time t. The event universe can be indicated as the Cartesian product: $\mathcal{E} = \mathcal{C} \times \mathcal{A} \times \mathcal{R} \times \mathcal{T}$.*

As stated in Definition 2, each event reports on the execution of an activity within a specific instance of the business process, usually called *case*.

Definition 2. *Case. Let $\mathcal{E}$ be a finite set of events. A case $\sigma \in \mathcal{E}*$ is a finite sequence of events belonging to $\mathcal{E}$ and related to a same process execution.*

All cases characterised by the same sequence of events are represented by the same trace, as stated in Definition 3.

Definition 3. *Trace. Let $\mathcal{A}$ be a finite set of activities. A trace $\theta \in \mathcal{A}*$ is a finite sequence of activities belonging to $\mathcal{A}$.*

Process Mining algorithms interpret an event log as a multi-set of traces and infer models comparing these sequences of events. We argue that this notion is not necessarily capturing the business goals of the organisation in addressing a case. For this reason, our pre-processing analysis starts by segmenting the event log base on business goals. This way, segments can be compared to relevant business requirements and further steps of refinement can be oriented based on a specific optimisation criterion. Given our definition of case, we can now formalise the definition of segment, according to Definition 4.

Definition 4. *Segment. Given n cases, a segment s is a union of cases: $s = (c_1 \cup c_2 \cup, \ldots, c_n)$ where s is a subset of an Event Log: $s \subseteq L$.*

A variety of criteria can be used to segment event logs [11], including temporal constraints, case type, business rule compliance, performance result, resources involved in the execution and others. In the rest of this paper we will generically refer to these criteria as business rules. From an operational point of view, a segment can be identified by a query over a set of predicates that can be joined with one of the elements composing an event, as proposed in [6]. Table 1 shows an excerpt of a real-life event log. Segmenting by the values in the field `Customer` we get three segments: $s_1 = \{Case1, Case5\}$, $s_2 = \{Case2, Case4, Case6\}$, and $s_3 = \{Case3\}$.

3.2 Step 2: Trace Profiling

Following [20], profiling a log can be described as the aggregation in a vector of a set of measures on the events composing a trace. These vectors can be used to calculate the distance between any two traces, using a suitable distance metric. In this work, we are proposing a new method for profiling traces that can be extended to segments and that offers a good trade-off between computational complexity and context aware encoding, as discussed in Sect. 5.

The basic idea is that the structure of the event log is reduced to a list of activities and each activity has a vector of positions. This vector is defined as a list of ordinal positions with corresponding frequency. With this definition, the set of elements representing an event is extended from a binary relation {case × activity} to a ternary relation {case × activity × position}, as stated in Definition 5. Nevertheless, the representation format is kept bi-dimensional by creating an element in the vector for each couple {activity × position}.

Definition 5. *Position profile. A position profile is a triple $apf = (a, p, f) \in \mathcal{E}$, denoting the occurrence of an activity a at the position p with the frequency f. The event universe can be indicated as the Cartesian product: $\mathcal{E} = \mathcal{A} \times \mathcal{P} \times \mathbb{N}$.*

As an example, we convert Table 1 adopting the definition above and obtaining the representation shown in Table 2. Activities a, b, h and i are always at a

Table 1. An example of real-life event log.

Case ID	Activity	Customer	Case ID	Activity	Customer
1	process creation	Gng inc.	4	process creation	MAS spa.
1	configuration manager	Gng inc.	4	configuration manager	MAS spa.
1	weight	Gng inc	4	me fabrication checker	MAS spa.
1	m_p	Gng inc.	4	weight	MAS spa.
1	stress	Gng inc.	4	stress	MAS spa.
1	me assembly checker	Gng inc.	4	m_p	MAS spa.
1	me fabrication checker	Gng inc.	4	me assembly checker	MAS spa.
1	design checker	Gng inc.	4	design checker	MAS spa.
1	design leader	Gng inc.	4	design leader	MAS spa.
2	process creation	MAS spa.	5	process creation	Gng inc.
2	configuration manager	MAS spa.	5	configuration manager	Gng inc.
2	me fabrication checker	MAS spa.	5	weight	Gng inc.
2	weight	MAS spa.	5	m_p	Gng inc.
2	stress	MAS spa.	5	me assembly checker	Gng inc.
2	m_p	MAS spa.	5	stress	Gng inc.
2	me assembly checker	MAS spa.	5	me fabrication checker	Gng inc.
2	design checker	MAS spa.	5	design checker	Gng inc.
2	design leader	MAS spa.	5	design leader	Gng inc.
3	process creation	Herw inc.	6	process creation	MAS spa.
3	configuration manager	Herw inc.	6	configuration manager	MAS spa.
3	weight	Herw inc.	6	me fabrication checker	MAS spa.
3	m_p	Herw inc.	6	weight	MAS spa.
3	me assembly checker	Herw inc.	6	stress	MAS spa.
3	stress	Herw inc.	6	m_p	MAS spa.
3	me fabrication checker	Herw inc.	6	me assembly checker	MAS spa.
3	design checker	Herw inc.	6	design checker	MAS spa.
3	design leader	Herw inc.	6	design leader	MAS spa.

fixed position, namely 1*st*, 2*nd*, 8*th*, and 9*th*. On the other hand, activity c is 3 times 3*rd* position and 3 times in 4*th* position. We call this table a *position profile*. More formally, a position profile can be encoded as an integer matrix via a two-dimensional function $f(x, y)$, where y is the temporal occurrence order of an activity x. The amplitude f of any pair (x, y) represents the number of occurrences of activity x at position y. Acyclic processes are represented by square binary matrices. In the case of processes containing cycles an activity can occurs in multiple positions[5].

---

[5] Clearly, by generating segments the information on the control-flow encoded in matrices is aggregated using a compensative approach that can bias the comparisons. We plan to address this problem in future studies by using intra- and inter-segment similarity metrics.

Table 2. Position profile of event log in Table 1 using a simplified view where letters a-i in the trace represent activities in the following order: ['Process Creation', 'Configuration Manager', 'Weight', 'M_P', 'Stress', 'ME Assembly Checker', 'ME Fabrication Checker', 'Design Checker', 'Design Leader']

Activity\position	p(1)	p(2)	p(3)	p(4)	p(5)	p(6)	p(7)	p(8)	p(9)
a	6	0	0	0	0	0	0	0	0
b	0	6	0	0	0	0	0	0	0
c	0	0	3	3	0	0	0	0	0
d	0	0	0	3	0	3	0	0	0
e	0	0	0	0	4	2	0	0	0
f	0	0	0	0	2	1	3	0	0
g	0	0	3	0	0	0	3	0	0
h	0	0	0	0	0	0	0	6	0
i	0	0	0	0	0	0	0	0	6

Our encoding of the event log as an integer matrix allows us to perform several types of distance analysis, from simple matrix distance to neighbourhood evaluation. In other words, our matrix offers a novel computation-friendly representation for business processes event logs.

3.3 Step 3: Compute a Similarity Measure

Based on the matrix introduced above we are able to compute a degree of similarity between two different matrices. This similarity can distinguish different traces, segments or event logs, leading to results that naturally encode the control flow of a trace. Thus, given two matrices A and B, the similarity function can be defined as a generic norm $n(A, B)$. For example, one could simply subtract the number of occurrences reported in matrix A from the one in B, or compute edit distance [5] or cosine distance [8], as discussed in Sect. 2. However, in this work we want to propose an original approach for comparing trace profiles. The motivating idea is to identify a method that is not biased by a specific probability distribution. As discussed in [23], cosine similarity and other common similarity metrics are designed to work with normal distribution only, while this assumption is not made explicit in most of the approaches that adopt them. We then decided to propose a method based on inferential statistics where hypothesis testing control the epistemiological assumptions.

Definition 6. *Hypothesis Testing. Let H_0 and H_{alt} denote the null and the alternative hypothesis respectively. Given two segments s_i and s_j, $\{s_i, s_j\} \subset s, \forall a \in \mathcal{A}$, a statistical test $ST(s_i, s_j, a)$ confirms H_0 when $\forall p$ from a it holds $(p_i, f_i) = (p_j, f_j)$; otherwise H_{alt} is confirmed.*

In Definition 6, ST is a statistical hypothesis test from parametric or non-parametric methods, a is an activity and p is its position, in accordance to

Definition 5. In our research, after some trial-and-error on various tests, we focused on Jensen-Shannon divergence test and on a group of non-parametric statistical two-sample hypothesis tests based on correlation, namely the Spearman's and Kendall's rank correlation coefficient. The results returned by correlation test are expressed in term of a *p-value*, or calculated probability, that is the probability of finding the observed, or more extreme, values when H_0 is true. To make a decision of either accept or reject null hypothesis we should define a preset value called significance level or α for estimating the p-value. If p-value $< \alpha$ then we have sufficient evidence to reject H_0 and H_{alt} may be accepted. Otherwise, if p-value $> \alpha$, there is not sufficient evidence to conclude that the H_{alt} may be correct[6].

The Jensen-Shannon divergence is closely related to the KullbackLeibler distance (KL) which in turn can be approximated by the classic Chi-Square test. Given two vectors $\mathbf{A}$ and $\mathbf{B}$, $KL(\mathbf{A}, \mathbf{B})$ is calculated as $\sum a_i \ln \frac{a_i}{b_i}$. The Jensen-Shannon divergence compares two vectors $\mathbf{V}$ and $\mathbf{U}$ by averaging their probability distributions in a new vector $\mathbf{M} = \frac{1}{2}(\mathbf{V} + \mathbf{U})$. For each vector, it is computed a pair of values describing this divergence through the pair $KL(\mathbf{V}, \mathbf{M})$ and $KL(\mathbf{U}, \mathbf{M})$. To obtain a final distance metric, it is required to average the resulted divergence values and re-size the final result computing the square root. We can formalise this as the formula in Eq. 1.

$$Jensen - Shannon\ Distance = \left(\frac{KL(\mathbf{V}, \mathbf{M}) + KL(\mathbf{U}, \mathbf{M})}{2} \right)^{\frac{1}{2}} \quad (1)$$

In Table 3 we compare two position profiles showing the results returned by the different metrics we considered. Note that for the Jensen-Shannon divergence the metric reports on the distance between two vectors, while when we use the p-values the metric reports about the probability that the two vectors were generated by distributions sharing same characteristics. We do not provide here a full explanation on how the Spearman's and Kendall's rank correlation coefficients were calculated. The interested reader can refer to [13] for details. To return an overall value about the comparison of the two segments we adopt different approaches in case the ST is returning a p-value or not. When ST returns a distance measure we simply average the results obtained for each activity (third to last column in Table 3). While dealing with p-values, we compute an index stating how many times the calculated probability is less than the significance level α (second to last and last columns in Table 3, taking $\alpha = 0.01$).

3.4   Step 4: Characterise Segments

We have now a measure of the dissimilarity level of two probability distributions of the activities' position in segments. Using this dissimilarity metrics we can

---

[6] Note that, when we do not reject H_0, it does not mean that H_0 is true. It means that the sample data have failed to provide sufficient evidence to cast serious doubt about the truthfulness of H_0.

cluster segments as illustrated in Fig. 4. An *assessment criterion* for validating a segment is obtained by imposing a minimum dissimilarity value in comparison to the others segments in the event log. Clearly, this procedure can be applied with different metrics looking for at least one metric where the criterion is met. Any segment that is not compliant with assessment criteria need to be re-sized. Moreover, our method provides us with a measure of the specific contribution that each activity has provided in characterising a segment. For example, using the Jensen-Shannon divergence test we can identify activity b, g, and i as those that are introducing most divergence. This information offer us an heuristic optimisation criterion for increasing or decreasing divergence.

If our interest is not particularly related to measuring a distance we can exploit the non-parametric tests that provide us with a measure of the correlation of two distributions. In modern use, "correlation" refers to a measure of a linear relationship between variables, while "measure of association" is usually referred to a measure of a monotone relationship between them. Two well-know examples that measure the latter type of relation are Kendall's tau and the Spearman rho metrics. Fig. 2 shows the difference between monotonic and non-monotonic relation. Differently to the Jensen-Shannon distance, these metrics tell us if two distributions have a similar trend, without measuring a precise distance on frequencies. We can, in other words, detect traces or segments that are similar because the shape of their probability distribution, even in presence of different absolute values.

Figure 3 displays two activities from Table 2 and provides an interpretation of the p-value of each segment. The red and blue line describe the distribution of these activities in segment 4 and 8, respectively. The pairwise comparison

Table 3. Table of position profiles of two segments and Jensen-Shannon distance (JS), Spearman's rank test (SR) and Kendall's rank test (KR)

activity \position	P1	P2	P3	P4	P5	P6	P7	P8	P9	P10	P11	JS	SR (p-value)	KR (p-value)
act a in s4	15914	377	0	0	0	0	0	0	0	0	0	0.0899	0.0061	0.00105
act a in s8	5418	0	0	0	0	0	0	0	0	0	0			
act b in s4	1	4959	0	18	2357	8	306	1	0	0	0	0.3752	0.0986	0.04297
act b in s8	0	4022	0	0	0	0	0	0	0	0	0			
act c in s4	0	0	1060	756	516	1002	690	308	187	121	0	0.2330	0.0000	0.00009
act c in s8	0	0	833	558	359	239	159	95	2	0	0			
act d in s4	0	130	671	830	712	870	833	509	245	164	0	0.2277	0.0008	0.00185
act d in s8	0	0	433	576	459	319	241	217	0	0	0			
act e in s4	0	36	314	491	653	892	761	998	414	385	2	0.2498	0.0000	0.00018
act e in s8	0	0	174	228	330	404	329	772	8	0	0			
act f in s4	0	29	323	382	526	959	1209	525	400	315	1	0.2472	0.0000	0.00010
act f in s8	0	0	207	299	408	481	528	320	2	0	0			
act g in s4	0	26	162	329	351	448	683	748	344	363	0	0.2817	0.0014	0.00137
act g in s8	0	0	129	264	325	464	569	491	3	0	0			
act h in s4	0	1	725	543	602	468	447	274	0	0	0	0.0860	0.0001	0.00052
act h in s5	0	0	469	320	364	337	419	336	0	0	0			
act i in s4	0	0	50	0	11	307	18	1272	1739	230	1332	0.5396	0.0165	0.00555
act i in s5	0	0	0	0	0	1	0	14	2198	0	0			
compare(s4, s8)												0.2592	0.88	0.88

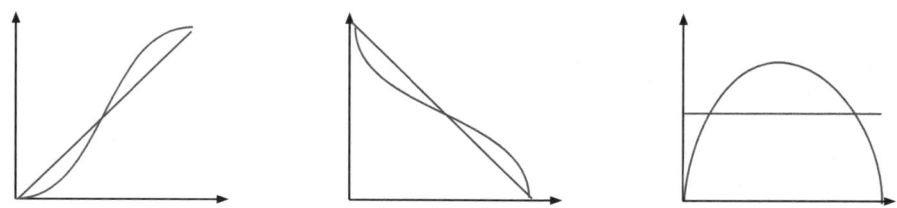

Fig. 2. A comparison between a monotonic and non-monotonic relationship (red color lines). From left to right: increasing monotonic, decreasing monotonic and non-monotonic relationship. (Color figure online)

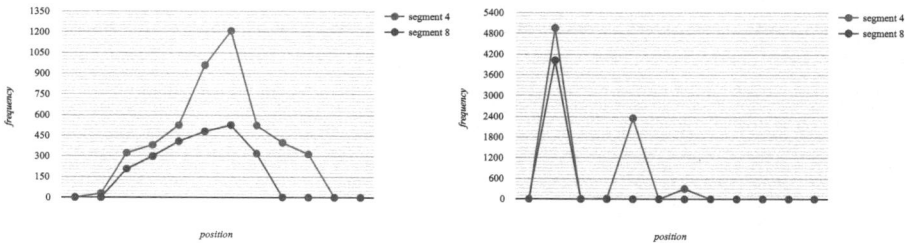

Fig. 3. Comparison between p-value of activities f(left) and b(right) in two different segments. In the picture on the left the probability distribution is monotonic, in the picture on the right it is not. (Color figure online)

of activities in the left figure shows a similar monotonic behavior. On the other hand, the right figure shows different behavior for the two activities. Even though both start with similar behavior, the 2nd halves of behavior are quite different. Indeed the applied test correctly assigns a lower p-value to the right figure.

4   Case Study

The method we described in the previous sections was applied to a real-life case study involving a manufacturing company in Italy. The event log collected by this company includes different business process related to product life-cycle management. Table 4 lists some descriptive statistics about this event log. The aim of the company was to discover real-life models that can be then used as a reference to identify cases that are deviating from the norm. In order to identify significant segments in the event log we considered Business Rules (BR) as criteria to construct segments. Each segment only includes cases consistent to a specific business rule. Then we used the Kendall's test as a metric for clustering segments.

A comparison of the results is shown in Fig. 4 where a dendrogram, or tree diagram, is used to illustrate the hierarchical arrangement of the clusters obtained. The thick red line in the figure helps to cut the dendrogram and returns the group

Table 4. Descriptive Statistics of the event log

#events	#cases	Mean case duration	Median case duration	Min duration	Max duration
94622	24858	61 h	7 s	0 mills	300 days

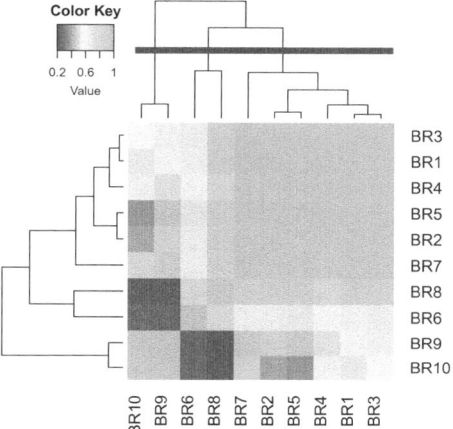

Fig. 4. Comparison table obtained with Kendall's Tau hypothesis test with $\alpha = 0.01$. The red line shows the cut line we applied for obtaining clusters compliant to the adopted assessment criteria. The clusters discovery were BR = (6,8), BR = (9,10) and BR = (1,2,3,4,5,7). Colours represent similarity values, as reported in the legend on top left size. (Color figure online)

of samples that belongs to the same cluster. By adjusting the assessment criterion, we can have more or less detailed group of segments in each cluster.

The next step is to perform process discovery for each cluster, as shown in Fig. 5. Significance of discovered model has been tested by asking to three managers of this company to rate in a Likert scale their agreement with the following sentence "*Do you think the model discovered improves your understanding of this business process?*". According to these managers, the models discovered using BR = (6,8) and BR = (9,10) are significant, corroborating with the discovered clusters[7]. On the other hand, the model discovered by BR = (1,2,3,4,5,7) did not improve their understanding of the business process, when compared to the model obtained from the entire event log[8]. Therefore, additional analysis is required for this cluster. Indeed, to fit the assessment criterion, quite a number of segments were included in this cluster, indicating that the applied segmentation was not really able to characterise a model.

---

[7] The rates provided are 3 "Neither agree nor disagree", 4 "Agree", and 4 "Agree".

[8] The rates provided are 1 "Strongly disagree", 2 "Disagree", and 2 "Disagree".

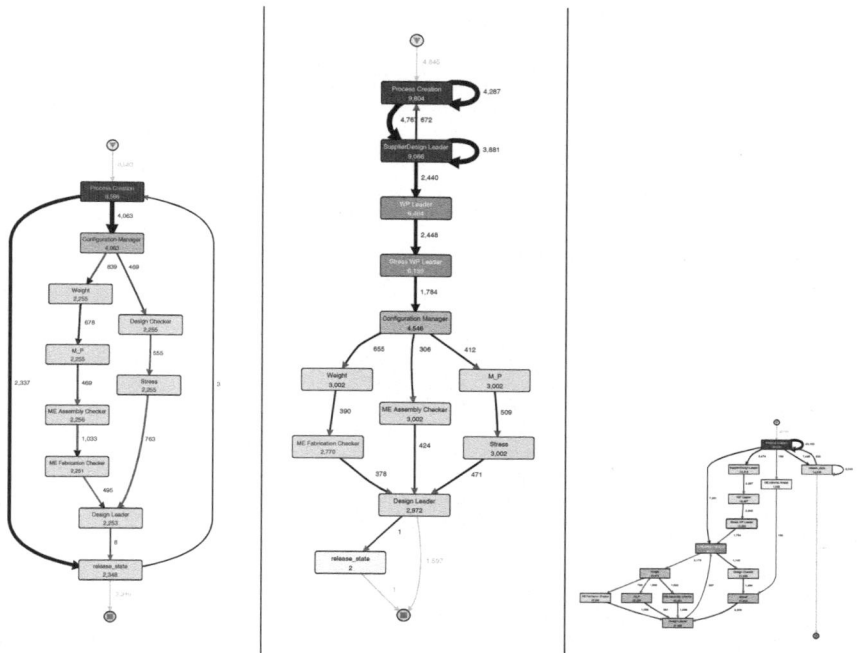

Fig. 5. From left to right, discovery model of BR = (6,8), BR = (9,10) and BR = (1,2,3,4,5,7).

5   Time Complexity Analysis

In this section we provide an evaluation of our method in terms of time complexity, and we compare it to other techniques available in the literature. As discussed in Sect. 2, naive solutions implies computational costs that are linear in the log size. In comparison to these solutions our approach has a higher complexity. Nevertheless, naive models do not account the trace structure [5] while our technique encode structural information in trace profiling. Indeed the time complexity we achieve is less than the one of other solutions taking into account the structure of the event log that, as already reported in Sect. 2, have to introduce some exponential factor.

In order to calculate the overall time complexity, we perform our analysis in three steps, so that it will be easier to understand. The evaluation of the proposed approach has focused on the contribution of statistical inference towards supporting the similarity of activities. In other words, the complexity of other involved techniques was not considered in our evaluation upon highlighting the main contribution. After identifying the complexity of the Jensen-Shannon divergence, we calculated the complexity of the Spearmans's rank and the Kendall's rank, our non-parametric hypothesis test algorithms.

5.1   Complexity of Jensen-Shannon Divergence Test

As mentioned in Sect. 3.3, the Jensen-Shannon calculation uses KL computation to obtain an initial segment distance. KL has $\mathcal{O}(a * p)$ where a is the number of activities and p is the possible positions acquired by activities. The other operations included in the Jensen-Shannon test have constant asymptotic complexity. This way, the final complexity is $\mathcal{O}(a * p + 1) = \mathcal{O}(a * p)$.

5.2   Complexity of Spearman's Rank Correlation Test

The complexity of Spearman's rank correlation test for two lists $x_1, \cdots, x_p$ and $y_1, \cdots, y_p$ is calculated as follows:

1. no tied ranks: $\rho = 1 - \left(\frac{6 \sum d^2}{p(p^2 - 1)} \right)$;

2. tied ranks: $\rho = \frac{\sum_i (x_i - \overline{x})(y_i - \overline{y})}{\sqrt{\sum_i (x_i - \overline{x})^2 \sum_i (y_i - \overline{y})^2}}$.

Where p is the maximum number of positions in a segment. The formula for no tied ranks has fewer operations than the tied rank formula; so we calculate only the complexity of the second formula. There are 2 averages, $2p$ differences, three sums with p summands and 1 division, 1 multiplication and 1 square root. Then the complexity will be $\mathcal{O}(2 + 2p + 3p + 1 + 1 + 1) = \mathcal{O}(p)$.

Before applying the formula, we need to sort the variables and obtain their ranks. Depending on the sorting algorithm, we can have different complexity. Best general sorting algorithms (such as Binary Tree Sort, Merge Sort, Heap Sort, Smooth Sort, Intro Sort, etc.) have the worst case complexity of $\mathcal{O}(p \, log(p))$. The overall complexity of Spearman's rank correlation test is the sum of the above steps which is $\mathcal{O}(p \, log(p)) + \mathcal{O}(p) = \mathcal{O}(p \, log(p))$.

5.3   Complexity of Kendall's Rank Correlation Test

In order to compute the number of concordance, discordance and ties, required to compute this test we need to compare each position with itself in a brute-force manner. If we consider all permutations of positions and eliminate comparison of position with itself, we obtain $\frac{p^2}{2} - p$ comparison which has the complexity of $\mathcal{O}(p^2)$.

In [15] were described sorting procedures that reduce this complexity. The basic idea is to sort the observation in one dimension and then sort this sorted values in the other dimension using a modified version of merge sort. This modified version takes advantage of having sorted values in the first dimension. As the complexity of merge sort of the algorithm in [16] is $\mathcal{O}(n \, log(n))$, the complexity of Kendall's rank correlation test can be reduced to $\mathcal{O}(p \, log(p))$.

6   Conclusion and Future Work

In this paper we described a novel method for improving the characterisation of event logs in preparation to PM. Our original contribution covers different aspects:

- We highlighted the different steps that must be integrated to work out a pre-processing task, underlining that a consistent representation of the different elements involved in these steps is required to support multiple iterations.
- We proposed a method for trace profiling that brings to trace clustering with linearithmic time complexity. Comparing it with approaches with higher complexity or similar complexity that require treatment over all activities in the log, we claim that our method can improve process modelling without increase the complexity of the computing effort.
- We proposed the adoption of distance metrics rooted in inferential statistics supporting explicit assumptions on the probability distribution that are used in tests or to get specific characterisation about the correlation between two distributions.
- We applied the proposed methodology in a case study to demonstrate its positive applicability.

Future work will develop along several avenues. On the one hand, the method we adopted to generate position profile can be refined to get additional sensitivity to structural information. For example, it is of interest to verify how duplicated or skipped activities impact on similarity measures. Moreover we have to experiment several clustering approaches testing the impact of different distance metrics on their performances. Furthermore, additional statistical tests can be considered, in particular for supporting multi-vector comparison. Finally, the assessment procedure can be enriched introducing maximisation criteria insisting on inter- and intra-cluster distances.

Acknowledgements. This work was partly supported by the project "Cloud-based Business Process Analysis" funded by the Abu Dhabi ICT Fund at EBTIC/Khalifa University.

References

1. Appice, A., Malerba, D.: A co-training strategy for multiple view clustering in process mining. IEEE Trans. Serv. Comput. **9**(6), 832–845 (2016)
2. Bernardi, S., Requeno, J.I., Joubert, C., Romeu, A.: A systematic approach for performance evaluation using process mining: the POSIDONIA operations case study. In: Proceedings of the 2nd International Workshop on Quality-Aware DevOps, pp. 24–29. ACM (2016)
3. Bogarín, A., Romero, C., Cerezo, R., Sánchez-Santillán, M.: Clustering for improving educational process mining. In: Proceedings of the Fourth International Conference on Learning Analytics And Knowledge, pp. 11–15. ACM (2014)
4. Bose, R.P.J.C., Mans, R.S., van der Aalst, W.M.P.: Wanna improve process mining results? In: 2013 IEEE Symposium on Computational Intelligence and Data Mining (CIDM), pp. 127–134. IEEE (2013)
5. Bose, R.P.J.C., van der Aalst, W.M.P.: Context aware trace clustering: towards improving process mining results. In: Proceedings of the 2009 SIAM International Conference on Data Mining, pp. 401–412. SIAM (2009)

6. Ceravolo, P., Azzini, A., Damiani, E., Lazoi, M., Marra, M., Corallo, A.: Translating process mining results into intelligible business information. In: Proceedings of the The 11th International Knowledge Management in Organizations Conference on The changing face of Knowledge Management Impacting Society, p. 14. ACM (2016)

7. Ceravolo, P., Fugazza, C., Leida, M.: Modeling semantics of business rules. In: Digital EcoSystems and Technologies Conference, DEST 2007, Inaugural IEEE-IES, pp. 171–176. IEEE (2007)

8. Cha, S.-H.: Comprehensive survey on distance/similarity measures between probability density functions. City **1**(2), 1 (2007)

9. Chen, J., Yan, Y., Liu, X., Yu, Y.: A method of process similarity measure based on task clustering abstraction. In: Ouyang, C., Jung, J.-Y. (eds.) AP-BPM 2014. LNBIP, vol. 181, pp. 89–102. Springer, Cham (2014). doi:10.1007/978-3-319-08222-6_7

10. Damiani, E., Ceravolo, P., Fugazza, C., Reed, K.: Representing and validating digital business processes. In: Filipe, J., Cordeiro, J. (eds.) WEBIST 2007. LNBIP, vol. 8, pp. 19–32. Springer, Heidelberg (2008). doi:10.1007/978-3-540-68262-2_2

11. de Leoni, M., van der Aalst, W.M.P., Dees, M.: A general process mining framework for correlating, predicting and clustering dynamic behavior based on event logs. Inf. Syst. **56**, 235–257 (2016)

12. Dumas, M., La Rosa, M., Mendling, J., Reijers, H.A.: Fundamentals of Business Process Management, vol. 1. Springer, Heidelberg (2013). doi:10.1007/978-3-642-33143-5

13. Gibbons, J.D., Chakraborti, S.: Nonparametric statistical inference. In: Lovric, M. (ed.) International Encyclopedia of Statistical Science, pp. 977–979. Springer, Heidelberg (2011). doi:10.1007/978-3-642-04898-2_420

14. Jain, A.K., Hong, L., Pankanti, S.: IEEE draft standard for XES - extensible event stream - for achieving interoperability in event logs and event streams. Technical report P1849, IEEE-SA (2016)

15. Joe, H.: Dependence Modeling with Copulas. CRC Press (2014)

16. Knight, W.R.:A computer method for calculating kendall's tau with ungrouped data. J. Am. Stat. Assoc. **61**(314), 436–439 (1966)

17. Luengo, D., Sepúlveda, M.: Applying clustering in process mining to find different versions of a business process that changes over time. In: Daniel, F., Barkaoui, K., Dustdar, S. (eds.) BPM 2011. LNBIP, vol. 99, pp. 153–158. Springer, Heidelberg (2012). doi:10.1007/978-3-642-28108-2_15

18. Rebuge, Á., Ferreira, D.R.: Business process analysis in healthcare environments: a methodology based on process mining. Inf. Syst. **37**(2), 99–116 (2012)

19. Rojas, E., Munoz-Gama, J., Sepúlveda, M., Capurro, D.: Process mining in healthcare: a literature review. J. Biomed. Inform. **61**, 224–236 (2016)

20. Song, M., Günther, C.W., van der Aalst, W.M.P.: Trace clustering in process mining. In: Ardagna, D., Mecella, M., Yang, J. (eds.) BPM 2008. LNBIP, vol. 17, pp. 109–120. Springer, Heidelberg (2009). doi:10.1007/978-3-642-00328-8_11

21. Van der Aalst, W.M.P.: Process Mining. Data Science in Action. Springer, Heidelberg (2016)

22. Dongen, B.F., Adriansyah, A.: Process mining: fuzzy clustering and performance visualization. In: Rinderle-Ma, S., Sadiq, S., Leymann, F. (eds.) BPM 2009. LNBIP, vol. 43, pp. 158–169. Springer, Heidelberg (2010). doi:10.1007/978-3-642-12186-9_15

23. Whissell, J.S., Clarke, C.L.A.: Effective measures for inter-document similarity. In: Proceedings of the 22nd ACM international conference on Information & Knowledge Management, pp. 1361–1370. ACM (2013)
24. Yoo, S., Cho, M., Kim, E., Kim, S., Sim, Y., Yoo, D., Hwang, H., Song, M.: Assessment of hospital processes using a process mining technique: outpatient process analysis at a tertiary hospital. Int. J. Med. Inform. **88**, 34–43 (2016)

A Taxonomy of Compliance Processes
for Business Process Compliance

Tobias Seyffarth$^{(\boxtimes)}$, Stephan Kühnel, and Stefan Sackmann

Martin Luther University Halle-Wittenberg, 06108 Halle (Saale), Germany
{tobias.seyffarth, stephan.kuhnel,
stefan.sackmann}@wiwi.uni-halle.de

Abstract. Dynamic markets and new technology developments lead to an increasing number of compliance requirements. Thus, affected business processes must be flexible and adaptable. Ensuring business processes compliance (BPC) is traditionally operationalized by means of controls, which can be described as simple target-performance comparisons. Since such controls are not always suitable for achieving BPC, the view is extended by so-called compliance processes. However, the definition and design of appropriate compliance processes for effective BPC depend on a multitude of process characteristics. To address this issue on a general level, we developed a taxonomy for compliance processes consisting of 9 dimensions and 37 characteristics. As a result, the taxonomy allows researchers and practitioners to classify compliance processes according to the state of the art in a formal way. Furthermore, it provides a systematic fundament for greater flexibility, i.e. an ad hoc integration of compliance processes into ongoing business processes to ensure BPC during runtime.

Keywords: Business Process Compliance · Classification · Compliance Process · Taxonomy

1 Introduction

Dynamic markets, competitive constraints, and technological developments require flexible business responsiveness as well as the flexible adaptation of affected business processes [1]. This includes adherence to prescribed and/or agreed-upon norms, which is known as compliance [2, 3]. Such norms can originate from various compliance sources, like laws, regulations or standards, which have to be interpreted and translated into numerous (organization-specific) compliance requirements [2, 4, 5]. Not only business scandals but also modern technological developments such as digitalization, big data, and cloud computing lead to new and changing norms, which define a constantly increasing number of compliance requirements [6, 7].

Usually, controls are used to ensure business process compliance (BPC) [8, 9]. In [10], a control activity is defined as a single target-performance comparison. This is a narrow definition, since not all compliance requirements can be operationalized accordingly (e.g. obligatory duties) [11]. Other authors describe controls as restraining or direct influence to enforce, observe, or verify compliance requirements [12]. In this context, methods for business process modeling can also be used for control modeling

© Springer International Publishing AG 2017
J. Carmona et al. (Eds.): BPM Forum 2017, LNBIP 297, pp. 71–87, 2017.
DOI: 10.1007/978-3-319-65015-9_5

[13–15]. Therefore, BPC approaches depict controls as reusable [4] and autonomous control processes [14, 15], i.e. combinations of one or more control activities. Thus, and due to the inconsistent understanding of the term "control", we rely on the term "compliance process" in this paper. We define a compliance process as an independent process (part) consisting of at least one compliance-related activity that ensures BPC.

Compared to backward compliance checking, which is subsequently analyzing log files, runtime compliance checking allows to instantly avoid or react to possible compliance violations during the business process execution [12, 16, 17]. Ensuring BPC by runtime compliance checking becomes a challenging task when taking process flexibility into account [4, 14, 15], and it becomes more challenging when process flexibility is understood as "flexibility by change" [18], i.e. a business process can be adjusted on a per-instance basis during its runtime [4, 14, 15]. The separate modeling of reusable compliance processes [4] and its ad hoc integration in ongoing business process instances is a promising approach for ensuring BPC during runtime [14, 15]. However, a major challenge is the determination of appropriate compliance processes, as they depend on a large number of different characteristics (c.f. [9, 13, 19–21]). The characteristics of a compliance process can determine its execution in a business process [22] or its efficiency and effectiveness (c.f. [9, 23]).

Meanwhile, a substantial body of research has discussed the characteristics of compliance processes. For example, Riesner and Pernul [21] classify compliance processes according to their security semantics, such as integrity or availability. Panko [24] distinguishes between detective, preventive, and corrective compliance processes. Gehrke [19] as well as Schultz and Radloff [13] make a distinction according to their timing, frequency, or nature. Nevertheless, none of the authors addresses process flexibility or the ad hoc integration of compliance processes into business processes. Furthermore, none describes the proposed classification in a comprehensible way. Thus, we address the following research question: what characteristics address an ad hoc integration of compliance processes, what are general characteristics of compliance processes and how can they be classified? To answer this question, we developed a comprehensive compliance process taxonomy according to the well-established approach of Nickerson et al. [25]. Within the taxonomy development, we conducted a structured literature review according to vom Brocke et al. [26] and Webster and Watson [27] to conceptualize the characteristics of compliance processes. The resulting compliance process taxonomy enhances the descriptive knowledge in the field of BPC with two main contributions [28]. First, it extends existing classifications [13, 19, 21, 24] according to characteristics that are relevant for the ad hoc integration of compliance processes in ongoing business processes to ensure BPC during runtime [22]. Second, it combines additional general characteristics of compliance processes (e.g. [9, 20, 29, 30]) in a traceable way.

The contribution is structured as follows: in Sect. 2, a compliance model is presented that provides the formal basis for an ad hoc integration of compliance processes in ongoing business processes. In Sect. 3, according to the approach of Nickerson et al. [25], the development of our taxonomy is described and made comprehensible. In Sect. 4, the resulting taxonomy is discussed in greater detail, and in Sect. 5, the taxonomy is evaluated, and an exemplary application is presented. Finally, a brief conclusion and research outlook are provided in Sect. 6.

2 Connecting Business Processes with Compliance Processes

To describe relevant elements and their interrelations in the field of BPC, compliance models are often used. Many compliance models display the connection between compliance requirements, business processes, and controls or control processes [2, 4, 5, 12, 22]. For an ad hoc integration of compliance processes in ongoing business processes, a compliance model must depict at least three major requirements: (a) a separate modelling of business processes and compliance processes [15]; (b) a detailed description of the connection between compliance requirements, business processes, and further compliance processes; and (c) a separated view of process scheme and process instance. To develop an adequate compliance model, we refer to prior research [22][1]. In Sect. 3, the compliance model is used to derive necessary characteristics of compliance processes. They are used within the taxonomy for the ad hoc integration of compliance processes in ongoing business processes.

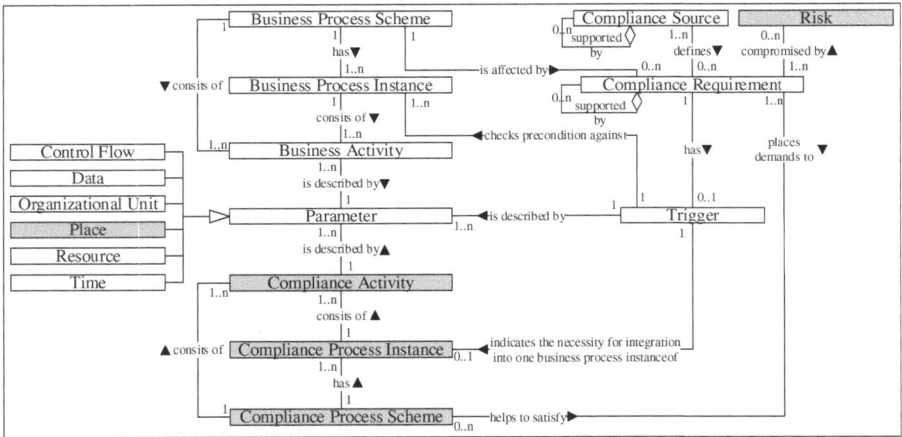

Fig. 1. Adjusted compliance model (based on [22])

At its core, the compliance model in [22] indicates the necessity for creation and integration of a compliance process instance into an ongoing business process instance. For example, a business process scheme called "Authorizing a Customer Loan Request" might be affected by a compliance requirement whenever a loan request exceeds a certain amount. Here, not every instance of the business process scheme is affected by the compliance requirement. Since the business and compliance processes are modelled separately, the compliance model in [22] aims at integration. Therefore, the connection between business activities (i.e. activities of the actual business process) and compliance process instances is formalized by parameters describing the properties

---

[1] Due to space limitations, we refer to [22] for a detailed explanation of the model.

of business activities and integration constraints of compliance process instances. A parameter consists of different parameter types, such as data, or organizational unit. Thus, the compliance model in [22] provides a method for determining all valid points within a business process at activity level for integrating compliance processes if required.

Although the view of the compliance model in [22] is already detailed, several adjustments are proposed. The resulting adjusted compliance model is visualized in Fig. 1 (adjusted and enhanced entities are highlighted in grey) and discussed in the following. The adjustments are grouped into three clusters, according to their characteristics, as follows: The original entities "control process scheme" and "control process instance" are redefined as "compliance process scheme" and "compliance process instance". As previously noted, a control activity is a single target-performance comparison [10] that is not capable of operationalizing all types of compliance requirements accordingly. Following our argument, a compliance process is the combination of one or more compliance activities that are capable of meeting an underlying compliance requirement. Thus, the entity "control activity" is also redefined as "compliance activity", representing an atomic work item (partly) ensuring that business processes are in accordance with a specified set of compliance requirements.

The compliance model is extended by an entity "risk" [4, 5]. Failures to meet compliance requirements increase the likelihood of risks materializing. For instance, a risk could occur as a result of an error in IT use. Usually, an error has an impact on the accuracy of financial reporting hence a compliance requirement could be compromised by a risk [9, 20, 29, 30]. This extension is required since the selection of a concrete compliance process from a set of possible alternatives is necessary in the context of flexible integration and economic risk is seen as a valid parameter for decision-making in the context of business processes [15].

The compliance model is extended by the parameter type "place" [31]. This is necessary, since the location of the execution of a business activity affects the integration of compliance processes in at least two ways: (a) depending on the place of processing different compliance requirements can affect a business process instance; and (b) the place might present integration constraints for compliance processes. For example, the German Federal Data Protection Act defines various compliance requirements for processing personal data [32] that are related to place, namely, place-related jurisdiction, which only affects business activities processed in Germany. An integration constraint occurs whenever a business or compliance process is constrained by its place of execution.

Based on this extended compliance model, it is still unclear how to categorize compliance processes (or their schemes and instances) in an appropriate way. Nevertheless, a classification is necessary for their appropriate selection and integration in ongoing business process instances. A taxonomy can be used to classify objects, thereby bringing order to the complex area of compliance processes [25, 33].

3   Taxonomy Development in Information Systems Research

Following the design science research (DSR) paradigm, our taxonomy is subsumed under the most fundamental artifact type "construct". Constructs define conceptual vocabulary providing the basis for the representation of problem domains or the construction of models [34]. According to Gregor and Mwilu et al. [28, 35], a taxonomy classifies objects or phenomena of interest, according to the dimensions that are relevant for characterizing and discriminating between these objects. As defined in Nickerson et al. [25], a taxonomy is used for the description and classification of existing or future objects in a specific domain. In addition, they define a taxonomy as a set of n dimensions each consisting of $k \geq 2$ mutually exclusive and collectively exhaustive characteristics. Mutually exclusive means that no object can have two different or even more than one characteristic in every dimension. Collectively exhaustive means that an object must have one characteristic in each dimension.

In the following, we refer to the definition of a taxonomy provided by Gregor and Mwilu et al. [28, 35] and apply the well-established methodological approach proposed by Nickerson et al. [25], which is variously used in the field of information systems [33, 35, 36]. According to Gregor and Mwilu et al., an object to be classified by the taxonomy can have different characteristics in one dimension. We will refer to this statement in the discussion of our taxonomy in Sect. 4.

The goal is to develop a "useful" taxonomy [25] for compliance processes and not a "correct" or the "best" one, since searching for the best solution is often intractable for information systems problems in DSR [34]. Figure 2 illustrates the applied approach for taxonomy development that is explained in greater detail in [25].

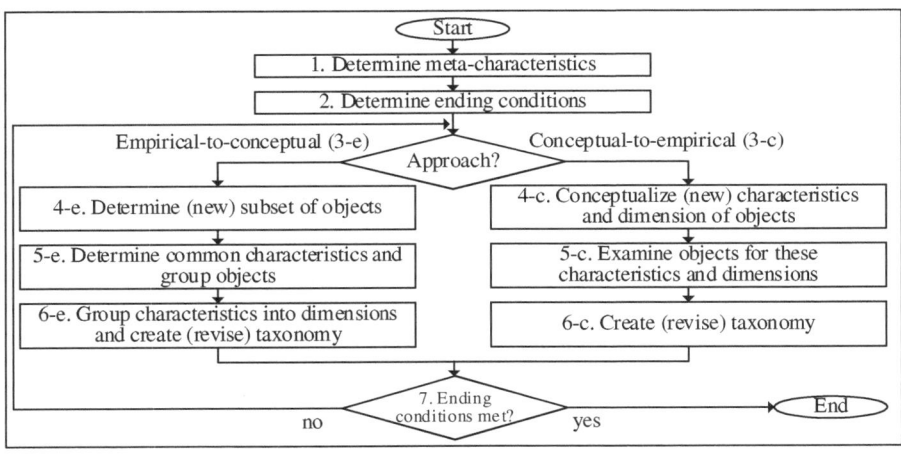

Fig. 2. Taxonomy development method [25]

Since comprehensibility is required in a scientific procedure, the first three steps of our taxonomy development are explained in detail. The last steps are explained by our application of the taxonomy development process.

Step 1. To begin with, meta-characteristics must be defined according to the purpose of the taxonomy [25]. The meta-characteristic is the most comprehensive characteristic that serves as the basis for choosing dimensions and characteristics. Since the aim is the specification of compliance processes that can be flexibly integrated in ongoing business processes, the following meta-characteristics are defined: (1) based on the compliance model as introduced in Sect. 2, the meta-characteristic "integration constraint" specifies integration constraints for a compliance process; (2) "modelling" specifies opportunities to model a compliance process at its compliance activity level, as well as opportunities for its integration in business processes; and (3) "property" specifies the properties of a compliance process for its processing.

Step 2. Defining a taxonomy involves an iterative approach; thus, ending conditions have to be defined [25]. Our development process stops if the number of dimensions allows the taxonomy to be meaningful (concise), and the dimensions as well as the characteristics provide differentiation among objects (robust). Furthermore, all dimensions of interest shall be identified (comprehensive) and new dimensions or characteristics shall be added easily (extendible). Finally, the development process ends if the taxonomy can explain the classified objects (explanatory).

Step 3. The actual taxonomy development begins with either an empirical-to-conceptual approach (step 3-e) or a conceptual-to-empirical approach (step 3-c). The decision regarding which approach shall be used depends on the availability of data. Step 3-e shall be used if significant data regarding the domain are available; step 3-c shall be used if the researcher has significant knowledge about the domain.

Because the field of BPC entails a large body of literature, the first iteration was done using the empirical-to-conceptual approach (step 3-e). Therefore, we conducted a structured literature review according to vom Brocke et al. [26]. As proper sources, we used the following databases: AIS Electronic Library (AISeL), EBSCOhost, IEEE Xplore Digital Library, the Journals of the American Accounting Association (AAA), and SpringerLink. Following Hevner et al. [34], the search was restricted to academic articles published within the last decade. We searched for contributions with full-text availability by using the search terms «application control», «("compliance process" OR "control process" OR "internal control") AND (category OR taxonomy)». As noted in [26], the resulting hits were selected by title, abstract, and full-text evaluation. Finally, [7, 13, 19, 21, 24] were identified as proper sources. According to Webster and Watson [27], a backward search was also conducted leading to [10, 37]. Furthermore, with respect to the domain, relevant international as well as German (available in English) standards and best practices in accounting, such as standards from the Institute of Public Auditors in Germany or COBIT and COSO [8, 9, 20, 29, 30, 38–41] were taken into consideration. A total of 16 highly relevant contributions were identified, providing the basis for our taxonomy development. Based on this body of literature, we defined the dimensions "controlled entity" and "assertion" within the meta-characteristic "integration constraints". Both dimensions set conditions according to the integration of a compliance process in a business process. Within the meta-characteristic "property", we defined the dimensions "timing", "type", and "execution". However, at this point, the taxonomy was not entirely concise according to the model proposed in Sect. 2.

Hence, the second iteration was conducted using the conceptual-to-empirical approach (step 3-c): the dimension "trigger" was added to the meta-characteristic "condition" and the dimensions "integration" and "compliance activity pattern" were added to the meta-characteristic "modelling". The dimension "integration" was identified due to extensive internal discussions. According to [22], a trigger is required in order to indicate the necessity to integrate a compliance process in an ongoing business process instance. According to [10], a control consists of a target-performance comparison and a deviation analysis. Each target-performance comparison should be supported by a recovery action to avoid compliance violations by handling negative results of the business process [37]. Therefore, we added the characteristics "target-performance comparison", "deviation analysis", and "recovery action" to the dimension "compliance activity pattern". By checking the ending conditions in step 7, the taxonomy was still not robust because the dimension "compliance activity pattern" does not provide a useful differentiation of compliance processes.

Therefore, the third iteration was conducted according to the conceptual-to-empirical approach. In practice, a compliance process does not exclusively consist of a deviation analysis or a recovery action. A compliance activity is always necessary for triggering a deviation analysis or a recovery action. Therefore, we divided the previously added dimension "compliance activity pattern" into the two dimensions "compliance requirement pattern" and "resolution pattern". In case of a negative target-performance comparison, the resolution pattern detects the results of a negative target-performance comparison and the recovery actions aim to avoid compliance violations.

The fourth iteration was conducted using the empirical-to-conceptual approach. We searched the International Standards on Accounting No. 315 [42] in order to build a more explanatory taxonomy. Including the results from the literature review of the first iteration, 15 different characteristics within the dimension "assertion" were identified. There are too many characteristics within one dimension for the application and comprehension of the taxonomy [25]. Therefore, the dimension was removed to provide a robust taxonomy according to [25]. In the next section, the resulting compliance process taxonomy and its characteristics are presented in detail.

4   A Compliance Process Taxonomy

Figure 3 shows the resulting compliance process taxonomy that finally allows for the categorization of compliance processes for ensuring BPC in an ongoing business process instance. The taxonomy consists of the three meta-characteristics "integration constraint" (dimension D 1 to D 3), "modelling" (D 4 to D 6) and "property" (D 7 to D 9).

D 1 – Trigger: As described in the compliance model, triggers are required for a flexible integration of compliance processes in ongoing business process instances [22]. A trigger is defined as a production rule that performs a certain action whenever a trigger condition is evaluated as true [43]. Following our understanding, a trigger

Dimension	Characteristics						
D 1: Trigger	D 1.1: Business Activity	D 1.2: Data	D 1.3: Organizational Unit	D 1.4: Place	D 1.5: Resource	D 1.6: Time	D 1.7: Frequency of Business Process Instance
D 2: Controlled Entity	D 2.1: Business Activity	D 2.2: Data	D 2.3: Organizational Unit	D 2.4: Place	D 2.5: Resource	D 2.6: Time	
D 3: Further Requirements for Execution	D 3.1: Data	D 3.2: Organizational Unit	D 3.3: Place	D 3.4: Resource	D 3.5: Time	D 3.6: No Further Requirements	
D 4: Compliance Requirement Pattern	D 4.1: Target Performance Comparison	D 4.2: Other Compliance Activity					
D 5: Resolution Pattern	D 5.1: Deviation Analysis	D 5.2: Recovery Action	D 5.3: No Resolution Pattern				
D 6: Integration	D 6.1: Sequential	D 6.2: Parallel	D 6.3: Start Event	D 6.4: End Event	D 6.5: Independent		
D 7: Timing	D 7.1: Preventive	D 7.2: Detective					
D 8: Type	D 8.1: Application Control	D 8.2: IT General Control	D 8.3: Business Control				Mutually exclusive characteristic
D 9: Execution	D 9.1: Automated	D 9.2: Manual	D 9.3: IT-dependent manual				Non-mutually exclusive characteristic

Fig. 3. Compliance process taxonomy

indicates the need to integrate a compliance process in a business process. The trigger condition is checked against each business activity and/or its properties.

In principle, the execution of a compliance process can be triggered based on an event or frequency [19, 37, 44]. For example, an event can be specified by the occurrence or absence of a business activity. Following the process flexibility type "flexibility by change" [18], the occurrence or absence of a business activity within a business process instance remains unclear at starting time. Therefore, the characteristic "business activity" (D 1.1) was added. Besides that, the relationships between further characteristics in D 1 are obvious [37]. A business activity may produce or consume "data" (D 1.2). An "organizational unit" (D 1.3) performs a business activity at a certain "place" (D 1.4) and may require several "resources" (D 1.5) to carry out the business activity. Finally the business activity has a processing time and starts and ends at a specific "time" (D 1.6). In the case of executing compliance processes for a random check of controlled entities, we added the characteristic "frequency of business process instance" (D 1.7). The characteristics in D 1 are defined as non-mutually exclusive [28, 35]. Therefore, a combination of them is possible to build the trigger.

D 2 – Controlled Entity: A controlled entity specifies the entity that is the subject of a compliance requirement. It arises from the entity parameter as discussed in the compliance model in Sect. 2. To integrate a compliance process in an ongoing business process, the controlled entity must be available. Therefore, the characteristic "business activity" (D 2.1) and its related elements "data" (D 2.2), "organizational unit" (D 2.3), "place" (D 2.4), "resource" (D 2.5), and "time" (D 2.6) were added. Like the characteristics in D 1, the characteristics in D 2 are defined as non-mutually exclusive [28, 35]. For instance, a compliance process, "Verify backup creation", is necessary to validate the occurrence of a required business activity, "Create data backup", at a certain place, "Data Center". Here, the compliance process has two characteristics within one dimension: the characteristic business activity is necessary to check the occurrence of the business activity "Create data backup", and the characteristic place is required to check the place of execution.

D 3 – Further Requirements for Execution: Sometimes, a compliance requirement induces further demands for the execution of a compliance process. Imagine that, a compliance requirement demands the authorization of a purchase request by an employee of the role manager. In this case, an organizational unit of the specified role must be available to execute the corresponding compliance process. For the description of these further requirements, again the characteristics "data" (D 3.1), "organizational unit" (D 3.2), "place" (D 3.3), "resource" (D 3.4), "time" (D 3.5), or even "no further requirement" (D 3.6) were added.

D 4 – Compliance Requirement Pattern: A pattern is an abstract process building that may contain various process elements [45]. Following this, a compliance requirement pattern is a pattern that contains various compliance process elements to ensure adherence to a compliance requirement [11].

A compliance process consists of at least one compliance activity directly enforcing compliance as well as optional activities resolving the results of former compliance activities or avoiding compliance violations. As stated in [14, 15], a control activity is simply a single "target-performance comparison" (D 4.1) returning a true/false statement. A target-performance comparison can be realized by various patterns, for example, by N-way match and plausibility or completeness checks [8, 9, 30, 37, 46]. Following COSO [8], a target-performance comparison can be used in so-called supervisory compliance activities to verify the correctness of other compliance processes or to validate compliance.

In some cases, satisfying compliance requirements does not need target-performance comparisons but rather non-comparative measures of the type "other compliance activity" (D 4.2). In the case of Section 14 of the German Banking Act, an approved loan amount greater than one million Euros requires the notification of the German Federal Financial Supervisory Authority [47]. The required notification by the credit institute can be realized by a compliance process consisting of patterns of the type "other compliance activity". Besides that, a compliance process sometimes consists of both types, which is why the characteristics are highlighted as non-mutually exclusive.

D 5 – Resolution Pattern: A resolution pattern comprises activities that analyze and react on the negative results of compliance requirement patterns. As noted, a target-performance comparison detects only deviations and finally returns only a true/false statement. Nevertheless, the identified deviation can be either positive or negative. In order to check this, the pattern "deviation analysis" (D 5.1) was added [10]. If the target-performance comparison returns as "false", the deviation analysis detects reasons for that result. Depending on the result of the deviation analysis and/or the degree of deviation, the handling of an affected business process instance can be determined. A "recovery action" (D 5.2) handles any further processing of affected business process instances to avoid compliance violations [8, 29, 37]. Recovery actions can also be interpreted as corrective control activities [8, 24, 41]. In [37], various recovery actions, such as "redo the affected business activity"; "in case of deviation, notify a responsible employee"; or "cancel the affected business process instance" are discussed. Besides that, sometimes a compliance process consists of neither a deviation analysis nor a recovery action [47]. Therefore, we also added the characteristic "no resolution pattern" (D 5.3).

D 6 – Integration: Based on the compliance model as proposed in Sect. 2, a compliance process has to be integrated in affected business process instances. In this context, the integration can occur as "sequential" (D 6.1) or "parallel" (D 6.2). As a result of the integration, the compliance process directly influences the business process instance and, e.g., its performance.

In addition, a compliance process is sometimes integrated in a business process only by its "start event" (D 6.3). As discussed above, the notification of the German Federal Financial Supervisory Authority does not require a confirmation of receipt. In this case, the result of the compliance process does not influence the business process instance. Both the business process instance and the integrated compliance process instance are completed separately. Moreover, only the "end event" (D 6.4) of a compliance process can be integrated into a business process instance. Finally, a compliance process can occur as completely "independent" (D 6.5) of a business process instance [9, 19]; for example the IT Governance Institute [9, 20] or ISACA [38] can demand a strategy for the cyclical backup of data and programs. A compliance process that satisfies this compliance requirement is performed independently of a special business process.

D 7 – Timing: As discussed in Sect. 2, our taxonomy considers compliance processes to realize BPC during runtime. In general, compliance processes are either "preventive" (D 7.1) or "detective" (D 7.2). A preventive compliance process attempts to keep deviations from occurring. In contrast, a detective compliance process attempts to uncover compliance violations in a business process instance after the time of their occurrence [8, 9, 12, 24, 39, 41]. The detection takes place before the ultimate objective of the business process instance has concluded [8]. In both cases, the critical part is the recovery action, which is used to correct or avoid an unintended event or result [8, 12, 17, 19, 24, 39]. Therefore, a detective compliance process may still (at least partly) enforce compliance [11].

D 8 – Type: Compliance processes are also categorized according to their type. There is a distinction between an "application control", an "IT general control", and a "business control" [20, 40, 46]. An application control (D 8.1) enforces, verifies, or observes compliance requirements within its embedded application. Furthermore, it observes the input, processing, and output of data processing [9, 29, 40, 46]. For instance, the identification and authentication of a user can be realized through a logical access control by checking a unique user ID and password [7, 29]. IT general controls (ITGC) (D 8.2) support the proper and continued operation of IT including application controls. ITGCs include compliance processes over program development, program changes, access to programs, or data and computer operations [7, 9, 20, 21, 40, 41]. They are either embedded or independent of business processes [9, 21]. A compliance process satisfying the illustrative compliance requirement "Implement a cyclical backup of data and programs" is categorized as an ITGC [9, 20, 29]. Finally, a business control (D 8.3) is a (manual) compliance process that is integrated in a business process or is independent from a business process [9, 40, 46].

D 9 – Execution: A compliance process is either processed "automated", "manual", or "IT-dependent manual". Within an automated (D 9.1) execution, all compliance activities of a compliance process are performed without human interaction. In contrast, a manual (D 9.2) compliance process is performed entirely without the use of any technology [8, 9, 12, 20, 29, 39, 41]. Hence, a compliance process of the type application control can also be processed manually [8, 29]. The design, implementation, and update of logical access controls are examples of a manual application control. If a compliance process consists of automatic and manual compliance activities, its execution is called IT-dependent manual (D 9.3) [9, 39]. In the next section, the evaluation of the developed compliance process taxonomy is discussed. Furthermore, the application is demonstrated by an example.

5 Compliance Process Taxonomy: Evaluation and Application

In DSR, artifacts are evaluated in two successive ways: ex ante or ex post [48, 49]. An ex-ante evaluation occurs prior to the artifact construction and focusses on artifact refinement during the design phase. An ex-post evaluation validates artifacts in use. Since the definition of a taxonomy is difficult to be evaluated ex post [25, 50], it was evaluated ex ante by conducting the following two steps: first, an extensive literature review based on a well-established methodology for literature review [26, 27] was conducted. Most of the taxonomy's dimensions and characteristics are derived from state-of-the-art literature. Second, our taxonomy was refined during the development phase [25].

After four iterations of the taxonomy development process, the (subjective) ending conditions were met. As required by [25], the resulting compliance process taxonomy is concise, robust, comprehensive, extendible, and explanatory. Consisting of nine dimensions, the taxonomy is concise. It is also robust with a maximum of seven characteristics in the dimension "trigger". By defining the meta-characteristics "integration constraint", "modelling", and "property" as well as the identification of dimensions and characteristics through the literature review, the taxonomy is also comprehensive. Furthermore, the taxonomy can easily be extended, e.g., by adding further characteristics to the dimensions "Compliance Requirement Pattern" or "Resolution Pattern". In addition, the taxonomy is explanatory, which is demonstrated by its application.

Figure 4 shows a simplified and adapted Purchase-to-Pay business process [51] that is modelled in BPMN 2.0 [52]. We call this process model an adapted process model because it contains business activities and integrated compliance processes. The business activities are modelled by the BPMN element "activity" (white); the integrated compliance processes are modelled by the BPMN element "collapsed sub-process" (grey). In the event of a negative result of the target-performance comparison the recovery action within each compliance process will terminate the adapted process

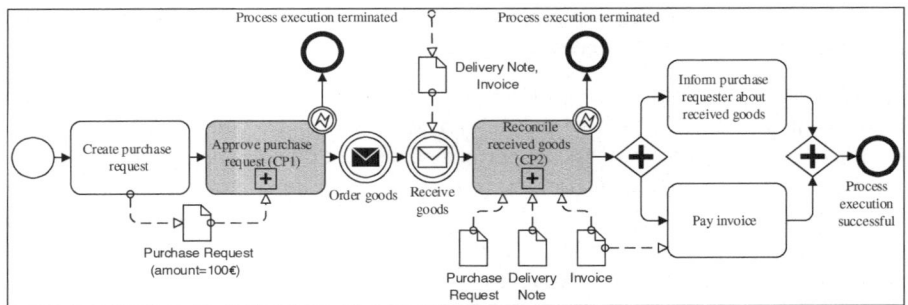

Fig. 4. Adapted purchase-to-pay business process model (based on [51])

instance. All required documents are also modelled by the corresponding BPMN element "data object".

The following two compliance requirements are assumed to affect the adapted purchase process discussed above:

- Compliance Requirement CR1: "Purchase requests with an order amount greater than 50€ must be approved according to existing order conditions by an employee of the role manager."
- Compliance Requirement CR2: "50% of all received goods must be reconciled to the purchase request."

Following the corresponding associations of the compliance model proposed in Sect. 2, we assume that the compliance process "Approve purchase request" (CP1) satisfies CR1. Equivalently, the compliance process "Reconcile received goods" (CP2) satisfies CR2. Figure 5 shows the compliance process taxonomy application by classifying both compliance processes.

	Dimension	Applicable characteristics of CP1	Applicable characteristics of CP2
Integration Constraint	D 1: Trigger	D 1.2: Data (amount of purchase request)	D 1.7: Frequency of Business Process Instance
	D 2: Controlled Entity	D 2.2: Data (purchase request)	D 2.2: Data (purchase request, delivery note, invoice)
	D 3: Further Require-ments for Execution	D 3.2: Organizational Unit (employee of the role manager)	D 3.6: No further Requirements
Modelling	D 4: Compliance Requirement Pattern	D 4.1: Target -Performance Comparison	D 4.1: Target -Performance Comparison
	D 5: Resolution Pattern	D 5.2: Recovery Action	D 5.1: Deviation Analysis D 5.2: Recovery Action
	D 6: Integration	D 6.1: Sequential	D 6.1: Sequential
Property	D 7: Timing	D 7.1: Preventive	D 7.2: Detective
	D 8: Type	D 8.1: Application Control	D 8.3: Business Control
	D 9: Execution	D 9.3: IT dependent manual	D 9.2: Manual

Fig. 5. Applied compliance process taxonomy

Based on CR1, CP1 is triggered by data because the amount of the purchase request is 100€, which is clearly above the 50€ stated in CR1. The controlled entity of CP1 is the purchase request. Therefore, the controlled entity is subsumed under the characteristic data. A further requirement for executing the compliance process is an organizational unit, since an employee of the role manager must approve the purchase request. CP1 consists of a target performance comparison. We assume that the purchase request is compared to requirements of internal guidelines. CP1 also contains a recovery action, which terminates the adapted process instance in the event of a negative target-performance comparison. CP1 is integrated sequentially in the business process and has preventive timing because the purchase request is checked before goods are ordered. Furthermore, CP1 is an application control that is performed as "IT-dependent manual" since the approval by an employee of the role manager still requires human interaction.

We assume that the purchase process instance occurs within the sample scope of the second compliance requirement. Furthermore, the compliance process is triggered by frequency of the business process instance. The compliance process CP2 compares the purchase request with the delivery note and invoice which means that it consists of a target-performance comparison. Thus, the controlled entity is of the type data. Besides that, there are no further requirements for the execution of CP2. We assume that CP2 performs a deviation analysis in case of a negative result of its target-performance comparison, before the recovery action takes place. The recovery action also terminates the adapted purchase process instance. Like the first compliance process, CP2 is integrated sequentially in the business process instance. In contrast to CP1, CP2 is a detective compliance process. The detection takes place after a possible violation (e.g. purchase request and delivery note do not match) but before the ultimate objective of the business process instance has occurred (i.e. informing the purchase requester about received goods). In addition, CP2 is a manually executed business control.

6   Conclusion and Outlook

Recent scandals and modern technological developments lead to new and changing norms, defining a constantly increasing number of compliance requirements [6, 7]. To comply with these requirements, the aim of business process compliance (BPC) is a comprehensive and comprehensible definition of compliance processes and their integration in business processes. For maintaining process flexibility in the context of BPC [18], a separate modelling of compliance and business processes as well as their ad hoc integration during runtime is discussed [4, 14, 15]. The determination of appropriate compliance processes for BPC becomes a major challenge, since they depend on a multitude of characteristics (c.f. [9, 13, 19, 20]). A major shortcoming of existing classifications for controls is the missing discussion of comprehensive compliance processes that are more than "simple" controls and their flexible integration in business processes. This research gap is addressed in the present contribution by the novel compliance process taxonomy which extends the descriptive DSR knowledge base. The taxonomy allows a classification of compliance processes based on 9 dimensions and 37 characteristics. Specifically, the first meta-characteristic "integration constraint" focuses on necessary characteristics to integrate compliance processes in ongoing business processes. The methodical development of the resulting compliance process taxonomy is considered to be comprehensive, concise, robust, extendible, and explanatory.

A well-known shortcoming of any literature review and taxonomy development is the fact that it is not possible to determine whether each and every relevant work, dimension, and characteristic has been found. However, by documenting the search for literature according to vom Brocke et al. [26] and Webster and Watson [27], as well as the use of the established methodology for taxonomy development by Nickerson et al. [25], comprehensibility in the development is provided in a scientific manner. Future research results, such as additional compliance requirement patterns might, thus, be incorporated relatively easily in the presented taxonomy.

The application of the developed compliance process taxonomy was demonstrated by a simplified business example. It shows that the taxonomy is an easy to-use tool for practitioners and academics. The taxonomy can also be used to focus further research or to provide a starting point for further investigations, e.g. by adding economic values for efficiency and effectiveness to several characteristics and, thus, to choose an optimal compliance process instance [23]. Furthermore, the taxonomy might be a sound basis for constructive discussions of or selections between alternative compliance processes by considering different process or execution types that satisfy the same compliance requirements. Another scenario could be the definition of different recovery actions depending on controlled entities or results of deviation analyses to enforce runtime compliance (at least partly).

References

1. Fdhila, W., Rinderle-Ma, S., Knuplesch, D., Reichert, M.: Change and compliance in collaborative processes. In: 12th IEEE International Conference on Services Computing (SCC 2015), pp. 162–169 (2015)
2. Sadiq, S., Governatori, G., Namiri, K.: Modeling control objectives for business process compliance. In: Alonso, G., Dadam, P., Rosemann, M. (eds.) BPM 2007. LNCS, vol. 4714, pp. 149–164. Springer, Heidelberg (2007). doi:10.1007/978-3-540-75183-0_12
3. Teubner, A., Feller, T.: Informationstechnologie, governance und compliance. Wirtsch. Inform. **50**, 400–407 (2008)
4. Schumm, D., Turetken, O., Kokash, N., Elgammal, A., Leymann, F., Heuvel, W.-J.: Business process compliance through reusable units of compliant processes. In: Daniel, F., Facca, F.M. (eds.) ICWE 2010. LNCS, vol. 6385, pp. 325–337. Springer, Heidelberg (2010). doi:10.1007/978-3-642-16985-4_29
5. Turetken, O., Elgammal, A., van den Heuvel, W.-J., Papazoglou, M.: Enforcing compliance on business processes through the use of patterns. In: 19th ECIS 2011 (2011)
6. Bagban, K., Nebot, R.: Governance und compliance im cloud computing. HMD **51**, 267–283 (2014)
7. Wallace, L., Lin, H., Cefaratti, M.A.: Information security and sarbanes-oxley compliance: an exploratory study. J. Inf. Syst. **25**, 185–211 (2011)

8. Committee of Sponsoring Organizations of the Treadway Commission (COSO): Internal Control - Integrated Framework. Framework and Appendices (2012)
9. IT Governance Institute (ITGI): IT Control Objectives for Sarbanes-Oxley, 2nd Edn. (2006)
10. Beeck, V., Wischermann, B.: Kontrolle. http://wirtschaftslexikon.gabler.de/Definition/kontrolle. html
11. Pretschner, A., Massacci, F., Hilty, M.: Usage control in service-oriented architectures. In: Lambrinoudakis, C., Pernul, G., Tjoa, A.M. (eds.) TrustBus 2007. LNCS, vol. 4657, pp. 83–93. Springer, Heidelberg (2007). doi:10.1007/978-3-540-74409-2_11
12. Turetken, O., Elgammal, A., van den Heuvel, W.-J., Papazoglou, M.P.: Capturing compliance requirements: a pattern-based approach. IEEE Softw. **29**, 28–36 (2012)
13. Schultz, M., Radloff, M.: Modeling concepts for internal controls in business processes – an empirically grounded extension of BPMN. In: Sadiq, S., Soffer, P., Völzer, H. (eds.) BPM 2014. LNCS, vol. 8659, pp. 184–199. Springer, Cham (2014). doi:10.1007/978-3-319-10172-9_12
14. Kittel, K., Sackmann, S., Göser, K.: Flexibility and compliance in workflow systems: the KitCom prototype. In: CAiSE Forum - 25th International Conference on Advanced Information Systems Engineering, pp. 154–160 (2013)
15. Sackmann, S., Kittel, K.: Flexible workflows and compliance: a solvable contradiction?! In: vom Brocke, J., Schmiedel, T. (eds.) BPM - Driving Innovation in a Digital World. MP, pp. 247–258. Springer, Cham (2015). doi:10.1007/978-3-319-14430-6_16
16. Kharbili, M., Medeiros, A., Stein, S., van der Aalst, W.M.P.: Business process compliance checking: current state and future challenges. In: MobIS (2008)
17. van der Aalst, W., van Hee, K., van der Werf, J.M., Kumar, A., Verdonk, M.: Conceptual model for online auditing. Decis. Supp. Syst. **50**, 636–647 (2011)
18. Schonenberg, M.H., Mans, R.S., Russell, N., Mulyar, N., van der Aalst, W.M.P.: Towards a taxonomy of process flexibility (extended version). BPM reports (2007)
19. Gehrke, N.: The ERP auditlab: a prototypical framework for evaluating enterprise resource planning system assurance. In: 43rd Hawaii International Conference on System Sciences (HICSS) (2010)
20. IT Governance Institute (ITGI): COBIT 4.1. Frameworks, Control Objectives, Management Guidlines, Maturity Models. Rolling Meadows (2007)
21. Riesner, M., Pernul, G.: Supporting compliance through enhancing internal control systems by conceptual business process security modeling. In: ACIS 2010 Proceedings (2010)
22. Seyffarth, T., Kühnel, S., Sackmann, S.: ConFlex: an ontology-based approach for the flexible integration of controls into business processes. In: Multikonferenz Wirtschaftsinformatik (MKWI) 2016, pp. 1341–1352 (2016)
23. Kühnel, S.: Toward a conceptual model for cost-effective business process compliance. In: Proceedings of the Informatik 2017. Lecture Notes in Informatics (LNI) (2017)
24. Panko, R.R.: Spreadsheets and Sarbanes-Oxley. Regulations, Risks, and Control Frameworks. Communications of the Association for Information Systems (2006)
25. Nickerson, R.C., Varshney, U., Muntermann, J.: A method for taxonomy development and its product service in information systems. Eur. J. Inf. Syst. **22**, 336–359 (2013)
26. Vom Brocke, J., Simons, A., Niehaves, B., Riemer, K., Plattfaut, R., Cleven, A.: Reconstructing the giant: on the importance of rigour in documenting the literature search process. In: 17th European Conference on Information Systems, pp. 2206–2217 (2009)
27. Webster, J., Watson, R.T.: Analyzing the past to prepare for the future: writing a literature review. MIS Quarterly **26**, 12–24 (2002)
28. Gregor, S.: The nature of theory in information systems. MIS Q. **30**, 611–642 (2006)

29. The Institut der Wirtschaftsprüfer in Deutschland e.V. [Institute of Public Auditors in Germany, Incorporated Association] (IDW) (ed.): Principles of Proper Accounting When Using Information Technology. IDW AcP FAIT 1 (2002)
30. The Institut der Wirtschaftsprüfer in Deutschland e.V. [Institute of Public Auditors in Germany, Incorporated Association] (IDW) (ed.): The Audit of Financial Statements in an Information Technology Environment. IDW AuS 330 (2002)
31. Tilburg University (ed.): COMPAS. Compliance-driven Models, Languages, and Architectures for Services. http://cordis.europa.eu/docs/projects/cnect/5/215175/080/deliverables/D2-1-State-of-the-art-for-compliance-languages.pdf
32. German Federal Ministry of Justice and Consumer Protection: Federal Data Protection Act (2009)
33. Silic, M., Back, A., Silic, D.: Taxonomy of technological risks of open source software in the enterprise adoption context. Inf. Comput. Secur. **23**, 570–583 (2015)
34. Hevner, A.R., March, S.T., Park, J., Ram, S.: Design science in information systems research. MIS Q. **28**, 75–105 (2004)
35. Mwilu, O.S., Prat, N., Comyn-Wattiau, I.: Taxonomy development for complex emerging technologies. The case of business intelligence and analytics on the cloud. In: 19th Pacific Asia Conference on Information Systems (PACIS 2015), pp. 1–16 (2015)
36. Glaser, F., Bezzenberger, L.: Beyond cryptocurrencies: a taxonomy of decentralized consensus systems. In: Proceedings of the ECIS (2015)
37. Namiri, K., Stojanovic, N.: Pattern-based design and validation of business process compliance. In: Meersman, R., Tari, Z. (eds.) OTM 2007. LNCS, vol. 4803, pp. 59–76. Springer, Heidelberg (2007). doi:10.1007/978-3-540-76848-7_6
38. ISACA (ed.): COBIT 5: A Business Framework for the Governance and Management of Enterprise IT. ISACA, Rolling Meadows (2012)
39. The Institute of Internal Auditors (IIA): SARBANES-OXLEY SECTION 404. A Guide for Management by Internal Controls Practitioners (2008)
40. The Institute of Internal Auditors (IIA): Global Technology Audit Guide (GTAG) 1. Information Technology Risk and Controls (2012)
41. The International Federation of Accountants (IFAC): ISA 315. Identifying and Assessing the Risks of Material Misstatement through Understanding the Entity and Its Environment (2009)
42. Public Company Accounting Oversight Board (PCAOB): Auditing Standard No. 5. An Audit of Internal Control Over Financial Reporting That is Integrated with an Audit of Financial Statements (2007)
43. Weigand, H., van den Heuvel, W.-J., Hiel, M.: Business policy compliance in service-oriented systems. Inf. Syst. **36**, 791–807 (2011)
44. Ramezani, E., Fahland, D., Aalst, W.M.P.: Where did i misbehave? Diagnostic information in compliance checking. In: Barros, A., Gal, A., Kindler, E. (eds.) BPM 2012. LNCS, vol. 7481, pp. 262–278. Springer, Heidelberg (2012). doi:10.1007/978-3-642-32885-5_21
45. Schäfer, T., Fettke, P., Loos, P.: Control patterns: bridging the gap between is controls and BPM. In: Proceedings of the 21st European Conference on Information Systems (ECIS), pp. 88–100 (2013)
46. Bellino, C., Wells, J., Hunt, S.: Auditing Application Controls. IIA, Altamonte Springs (2007)
47. German Federal Financial Supervisory Authority: Banking Act of the Federal Republic of Germany (Kreditwesengesetz, KWG). KWG (2016)
48. Pries-Heje, J., Baskerville, R., Venable, J.R.: Strategies for design science research evaluation. In: ECIS 2008 Proceedings (2008)

49. Sonnenberg, C., Brocke, J.: Evaluations in the science of the artificial – reconsidering the build-evaluate pattern in design science research. In: Peffers, K., Rothenberger, M., Kuechler, B. (eds.) DESRIST 2012. LNCS, vol. 7286, pp. 381–397. Springer, Heidelberg (2012). doi:10.1007/978-3-642-29863-9_28

50. Tremblay, M.C., Hevner, A.R., Berndt, D.J.: Focus Groups for Artifact Refinement and Evaluation in Design Research. Communications of the Association for Information Systems 26 (2010)

51. Namiri, K.: Model-Driven Management of Internal Controls for Business Process Compliance. Karlsruhe (2008)

52. OMG (ed.): Business Process Model and Notation (BPMN). http://www.omg.org/spec/BPMN/2.0/PDF/

Improving Pattern Detection in Healthcare Process Mining Using an Interval-Based Event Selection Method

Amirah Alharbi[1,2]([envelope]), Andy Bulpitt[1], and Owen Johnson[1]

[1] School of Computing, University of Leeds, Leeds, UK
mll3ama@leeds.ac.uk
[2] Computer Science Department, Umm Al-Qura University,
Makkah, Kingdom of Saudi Arabia

Abstract. Clinical pathways are highly variable and although many patients may follow similar pathway each individual will experience a unique set of events, for example with multiple repeated activities or varied sequences of activities. Process mining techniques are able to discover generalizable pathways based on data mining of event logs but using process mining techniques on a raw clinical pathway data to discover underlying healthcare processes is challenging due to this high variability. This paper involves two main contributions to healthcare process mining. The first contribution is developing a novel approach for event selection and outlier removing in order to improve pattern detection and thus representational quality. The second contribution is to demonstrate a new open access medical dataset, the MIMIC-III (Medical Information Mart for Intensive Care) database, which has not been used in process mining publications.

In this paper, we developed a new method for variations reduction in clinical pathways data. Variation can result from outlier events that prevent capturing clear patterns. Our approach targets the behavior of repeated activities. It uses interval-based patterns to determine outlier threshold based on the time of events occurring and the distinctive attribute of observed events.

The approach is tested on clinical pathways data for diabetes patients with congestive heart failure extracted from the MIMIC-III medical database and analyzed using the ProM process mining tool. The method has improved model precision conformance without reducing model fitness. We were able to reduce the number of events while making sure the mainstream patterns were unaffected. We found that some activity types had a large number of outlier events whereas other activities had a relatively few. The interval-based event selection method has the potential of improve process visualization. This approach is undergoing implementation as an event log enhancement technique in the ProM tool.

Keywords: Process mining · Healthcare processes · Interval pattern · Variation reduction · Feature selection · MIMIC-III · Event log quality

© Springer International Publishing AG 2017
J. Carmona et al. (Eds.): BPM Forum 2017, LNBIP 297, pp. 88–105, 2017.
DOI: 10.1007/978-3-319-65015-9_6

1 Introduction

Process mining aims to construct a model of business process using event logs extracted from business information systems and a process discovery algorithm [1] implemented in software tools such as the ProM framework [2]. There is growing interest in using process mining on data from electronic health record systems to model and improve care processes and to reduce costs [3] despite widely recognized issues with data quality [4].

Event logs in electronic health record systems have a considerable amount of variation which can hinder process model discovery. Event log preprocessing is a critical step for process mining research and this is recognized in the 2011 Process Mining Manifesto [5] as the first challenge for process mining. Outliers' events can be defined as events that prevent capturing clear patterns; such events affect the quality of process mining efforts. There are different issues related to event log quality such as missing events, imprecise timestamps and repeated events [6]. Repeated events, or duplicate tasks, occur when the same activity has been executed multiple times in the same case. In critical care, for example, the incidence of repeated events is high because events include periodic monitoring (known as "charting") of heart rate, blood pressure and other vital signs.

From a process mining point of view, repeated activity is a significant confounding factor that can prevent generating useful models [7]. Typically, the handling of frequently repeated tasks has been addressed in a model discovery phase [8–10] however, most current methods are tied to specific process discovery algorithms which restrict more general use.

Dealing with repeated activity as a preprocessing step has received relatively little attention in the process mining community. Moreover, to the best of our knowledge, no existing work has tackled variation reduction of repeated activity using activity temporal patterns. Although there are around 20 plugins in the ProM (version 6.5) process mining tool for log preparation, only two filters can be used for filtering repeated activity preparation. These filters are called *merge subsequent events* and *remove event type* [11]. They help to reduce the number of events however, no attention is paid for preserving time information about merged/removed events.

In this paper, we aim to present a new approach to filtering the outlier event of repeated activity using an interval-based selection method as a preprocessing step to applying process mining discovery techniques as a reusable method. This method aims to reduce the number of repeated events with more attention for preserving the mainstream temporal pattern. The method uses the interval-pattern of repeated activity as a threshold to remove outlier events.

The remainder of this paper is organized as follows: related work on handling duplicate activity as a pre-mining step is discussed. Section 3 describes the MIMIC-III medical dataset and outlines the healthcare data model used for event log extraction. Section 4 demonstrates two fundamental steps of event log preparations in order to provide a baseline event log that works as input for our interval-based event selection method. In Sect. 5, we present an analysis of our approach. A controlled evaluation with an existing technique is conducted along with the explanation of the impact of this approach on model precision and representational quality. In Sect. 6, the conclusion and future work is discussed.

2 Related Work

Process mining aims to construct a process model using an event log and a process discovery algorithm [1]. It has been applied in healthcare to improve care process and to reduce costs [3].

A few papers in the process mining literature have addressed repeated activity as a preprocessing step. In [12], the problem of repeating tasks was addressed by refining activity labels in a preprocess stage. This solution labeled repeated activity based on its context for instance, 'payment' activity can occur at the start of a process instance or at the end. Although this approach adopted accurate steps for detecting repeated activity, the method is not applicable in the case of large amount of repeated activities, such as those we found in healthcare data, because it increases the number of distinct activities.

Two papers [11, 13] have mentioned the idea of merging repeated events into one single event. This approach is implemented in ProM as an event log enhancement filter named merge subsequent events. It aims to merge consecutive events of the same activity. The merge subsequent events filter has three options of merging which are (1) merge by keeping the first event, (2) merge by keeping the last event or (3) merge by considering the first as start time and the last as end time. Using this method helps to reduce the number of events however, there are a number of limitations to be discussed. The first and second options of merging ignore the time aspect between events and concentrate on reducing the number of events at the cost of losing time information. The third type of merging may result in misleading event duration. In this paper, our aim is to improve on these tools to address the specific challenges of remove outliers in healthcare event log and compare our new method to the available techniques.

3 MIMIC-III Database

MIMIC-III (Medical Information Mart for Intensive Care) [14] is a publicly available medical research database of de-identified records of patients who were admitted to the Beth Israel Deaconess Medical Centre in Boston, USA between 2001 and 2012.

MIMIC-III database is integrated from multiple sources which include the hospital electronic health records, social security administration death master file and two distinct critical information systems that are called Philips CareVue and iMDSoft Metavision. The different data structures between the two critical information systems used by the hospital have largely been resolved at database integration stage. It is an important medical database that provides free access to researchers under agreement licenses which prohibit any attempt to re-identify patients. Different types of medical data are available, such as readings of vital signs, medications, laboratory tests, nurses' and physicians' observations and notes, fluid balance, diagnosis and treatments codes, care giver information, length of stay and time of death.

The data comprise 58,976 hospital admissions, and 46,520 distinct patients. 55.9% of the patients are male and 44.1% are female. There are around 380 types of laboratory measurements and 4,579 types of Intensive Care Unit (ICU) "charted" observations, such as heart rate and blood pressure. The admissions cover five critical care units which are the Coronary Care Unit (CCU), Cardiac Surgery Recovery Unit (CSRU),

Medical Intensive Care Unit (MICU), Surgical Intensive Care Unit (SICU) and Trauma Surgical Intensive Care Unit (TSICU).

The MIMIC dataset has been used in 134 publications mostly describing data mining and machine learning approaches [15]. None of these have described a process mining approach. In this paper, we describe how we have used the MIMIC-III database to extract and process mined an event log in order to explore patients' pathways for diabetes patients as a precursor to further work in diabetes. Our clinical advice is that these patients can be expected to have complex medical histories and complex care pathway patterns.

3.1 MIMIC-III and Process Mining

MIMIC-III can be used as a rich data source for process mining applications because it has many records with timestamps that can be extracted as medical events. There are 16 tables out of 26 tables in MIMIC-III database that contain medical events. These tables are used as a healthcare data model, which is discussed in the following section, for our healthcare process mining research.

In order to respect patient confidentiality the MIMIC-III dataset de-identification process included obfuscation of dates. The dates of all events have been shifted into the future using time offsets randomly generated for each patient. This approach preserves the time intervals and ensures the sequence of medical events are internally consistent but it means that certain process mining analytics approaches such as looking for arrival time bottlenecks cannot be used.

There are two main data types for time attributes in MIMIC-III which are chart time and chart date. They provide different time resolution of the event for instance, the chart date field has date only without time, this is because the accurate time for that event is not known, whereas chart time field has date and time with hour, minute and second of that event.

Most of chart time fields are recorded in the database with two columns, store time and chart time. In healthcare processes, observations are usually charted and then validated by a care giver such as a nurse. The validation process usually happens within an hour [14]. Therefore, chart time is the time when an observation is charted while store time is the time when the observation is validated. In the scope of this paper, we use chart time as the event time because it is the closest to reality. In related work we prove a structured assessment of the data quality issues related to process mining of MIMIC-III [16].

3.2 The Healthcare Data Model

A healthcare data model is a model that shows the relation between tables in a medical database that may contain healthcare events. The data model is significant in process mining research because it helps to extract event logs and to understand process oriented questions [3]. We developed a healthcare data model by analyzing the MIMIC-III database and using table descriptions based on [14, 15]. Figure 1 shows the Entity-Relationship

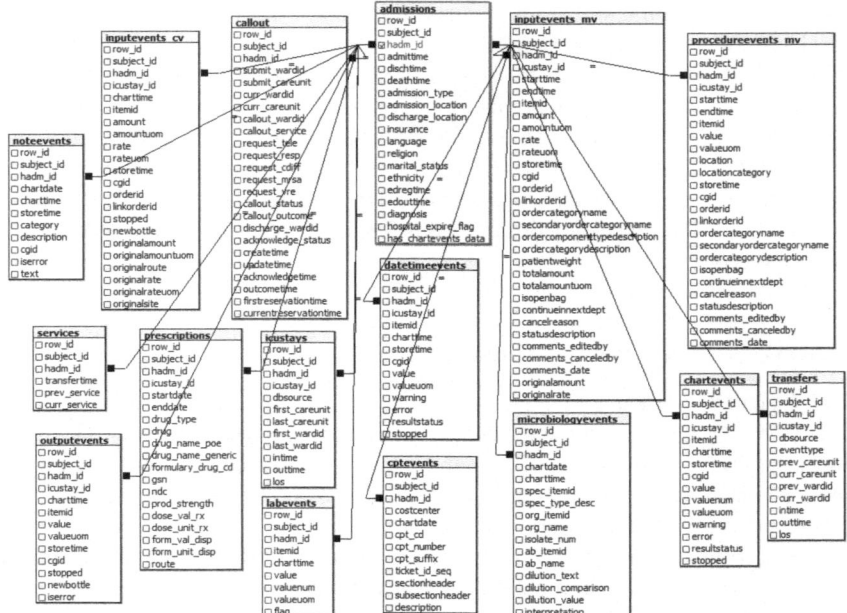

Fig. 1. E-R diagram of MIMIC-III data model constructed using PostgreSQL.

(E-R) diagram we constructed for the MIMIC-III database. The MIMIC-III data model is in effect a subset of the healthcare reference model that is discussed in details in [3], however is limited to data related to hospital patients with intensive care admissions.

The relevant healthcare events in our data model can be categorized into six groups of events which are administrative events, charted events, test events, medication events, billing events and report events.

In the following section, a description of the six event categories is provided along with a brief description of the sourced tables, for more detail about the tables the reader may refer to [14].

1. **Administrative events** identify patients' admission pathways which show if a patient has been admitted from emergency department or the patient has a pre-arranged admission. Also, administrative events include all patient transportation activities during their stay in different care units of the hospital through to a discharge event. This group of events is located in *Admissions, Callout, Transfer, ICU* stay tables.

Admissions table: holds demographic information about the patient, admission time, emergency department (ED) registration time '*edreg*', emergency department out time '*edout*', discharge and death time, discharge and death time.

Callout table: contains information about the time of discharge request and the time of the request outcome if it is fulfilled or cancelled.

Transfer table: holds information about patient transportation such as the time when a patient is moved in or moved out of different wards which include different critical care units.

ICU stay table: this is a sub-table from *Transfer* table especially for patients' transportations in Intensive Care Units (ICUs).

2. **Charted events** contain all bedside observations that are related to vital signs measurements such as heart rate and blood pressure or other intervention activity. This group of events is stored in the *Chart-events* and *Date-time-events* tables.

Chart-events table: has all patients' charted observations. There are more than 4500 types of charted observation. The table includes information about the time when an observation is taken and the time of observation validation performed by clinical staff.

Date-time-events table: this table contains the observation date of particular interventions such as dialysis or insertion of lines.

3. **Test events** correspond to all tests that have been measured on the patient such as laboratory tests and test results. This category of events is captured in *Output-events, Microbiology-events* and *Lab-events* tables.

Output-event table: has all output measurements for example, urine or blood. This table stores the time and value of the output measurement when is taken from the patient.

Microbiology-events table: this table contains information about tests and antibiotic sensitivities.

Lab-events table: this table has around 380 items for measurements some of them related to hematology and chemistry. It records output and microbiology results.

4. **Medication events** include prescribed medication and intravenous medication. These events can be extracted from *Prescription* and *Input-events-CV* and *Input-events-MV* tables.

Prescription table: this table contains information about when a drug starts and ends besides prescription order if it is needed.

Input-events-CV and *Input-events-MV* tables: these tables are generated from different healthcare information systems (CareVue and Metavision) but both contain information about the time when a medication intake occurred, for example enteral feeding, is recorded and its value. Some more transactional events are supported by *Input-events-MV* table such as the time when intake is ended or an intake order is updated

5. **Billing events** contain a list of medical procedures that are performed on patients that are used for billing services. Billing events can be extracted from *CPT-events* table.

CPT-events table: this table has a list of Current Procedural Terminology (CPT) codes for medical billing purposes. It contains information that shows the time of performed procedures.

6. **Report events** include different types of reports such as nurse notes and radiology notes. Report events are captured in the *Note-event* table.

Note-event table: this table has information about different types of notes, the date of reported notes and the ID of the caregiver who reported it.

It should be noted that, these events are distributed in various tables however, all tables have the basic requirements of process mining such as, a unique subject id, which corresponds to patient id, and a unique admission id, event, event time, some event attributes and some resources are associated with events which can be generated from the care-givers table. Table 1 provides a summary of process mining principle components in MIMIC-III.

Table 1. Process mining principle components in MIMIC-III

Table	Has timestamp		Has duration	Has observed item id[a]	Has care giver	Has cost
	Time and date	Date only				
Admissions	Yes		Yes	Yes	No	Yes
Chart-events	Yes		No	Yes	Yes	No
Input-CV	Yes		No	Yes	Yes	Yes
Input-MV	Yes		Yes	Yes	Yes	Yes
Output	Yes		No	Yes	Yes	Yes
Lab-events	Yes		No	Yes	Yes	Yes
Prescription	Yes		Yes	Yes	No	No
Note-events		Yes	No	No	Yes	Yes
Call	Yes		Yes	No	No	No
Cpt-event		Yes	No	Yes	No	Yes
Procedure MV	Yes		Yes	Yes	Yes	Yes
Transfer	Yes		Yes	No	No	Yes
ICU stay	Yes		Yes	Yes	No	Yes
Date-time	Yes		No	Yes	Yes	Yes
Microbiology	Yes	Yes	No	Yes	No	No

[a]Item id is one example of many event attributes can be extracted. It will be used in Sect. 5.

3.3   Extracting Event Logs from MIMIC-III

Although many modern business information systems automatically generate event logs, there are some information systems, including electronic health records that store process activities implicitly and consequently need a method for event log extraction. MIMIC-III is an object-relational database that is built using a PostgreSQL database management system. It does not support automatic extraction of event logs and we have therefore extracted the event log manually using SQL queries. The healthcare data model is used to guide event log extraction. In this paper, we used diabetes patients with congestive heart failure (CHF) as a use case.

```
SELECT subject_id, hadm_id, activity, time, cgid, cost
FROM Patients_Events // Patients_Events is a created table that contains all diabetes patients
activities where each row corresponds to one event
WHERE
    Patients_Events.hadm_id IN
        ( SELECT hadm_id
        FROM
        mimiciii.admissions
        WHERE
        diagnosis like '%CONGESTIVE HEART FAILURE' or
        diagnosis like '% congestive heart failure');
```

Fig. 2. An example of SQL query that is used to extract event logs

An example of SQL query that is used to extract an event log is shown in Fig. 2. The summary statistics of the extracted event log is shown in Table 2. There are 296 distinct admissions for 264 patients. Also, there are more than 2,300 activity types and more than 1,900,000 events which correspond to activity instances. The pathway variation reaches 100% among patients which means no common pathway is found. Admission id, hadm_id, is used as the case id in all our experiments.

Table 2. Summary of the extracted diabetes event log

Pathway characteristics	
Admissions (cases)	296
Patients	264
Patients with readmission	25
Variations	100%
Activity	$\sim 2,300$
Events	$\sim 1,900,000$
Mean event per case	$\sim 7,000$
Minimum event per case	55
Maximum event per case	$\sim 71,200$

4 Baseline Event Log Preparation

In this section, we demonstrate two fundamental steps of event log preparation in order to provide a baseline event log that works as input for our interval-based event selection method. Taking into account the statistics of the extracted event log as shown in Table 2, these two steps are crucial for managing event log quality.

4.1 Event Log Processing Step 1: Solve Batch Events

In the MIMIC-III database, there are some data quality issues such as missing accurate timestamps which is the result of batched events. Batch processing is the execution of several events at once and recording them with the same time, for example a group of laboratory results received at the same time. The issue of batch processing also leads to a huge number of fine-grain events that increase process model complexity. In our data model, the tables *Chart-events* and *Lab-events* contain a large number of batch events which should be addressed as a preliminary step for mining patient pathways.

Each patient in the ICU has been checked on a regular basis at varying intervals. The different measurements that are taken in each check have been recorded with the same time. For process mining purposes we are focusing on the process of charted observations regardless of which items are checked therefore all items are consolidated into a single charted event. Our hypothesis is that handling batched events as a single event simplifies the process model and improves process mining quality.

We re-extracted batched events with the same event label. The extraction includes tables that have batched events such as *chart-event* and *lab-event*. More precisely, for different chart measurements in the *chart-event* table such as Calcium, Glucose and Platelet count are all extracted under the name of Chartevent activity.

Results

This method has significantly reduced the number of activity types and the number of events which in turn reduced model complexity. It should be noted that, reducing the number of activities using this method does not lead to significant information loss because from a process mining perspective the exact name of measurements in the ICU is less important when we aim to mine the abstracted process model. We are able to capture the events occurred in *chart-event* and *lab-event* tables. Table 3 shows pathway characteristics after applying this manipulation.

Table 3. Summary of the extracted diabetes event log after processing step 1

Pathway characteristics	
Admissions (cases)	296
Patients	264
Variations	100%
Activity	35
Events	252,454
Mean event per case	853
Minimum event per case	28
Maximum event per case	10639

Although this method reduces the number of activities and events, the variation of patients' pathways is still extremely high and the event log needs further manipulations.

4.2 Event Log Processing Step 2: Mapping Fine-Grained Activities into Main Activity

In MIMIC-III there are two categories of fine-grained activities. The first category is transactional events and the second category is ontological events. The transactional event is an event that provides information about the activity - when it starts, updates, comments and finishes. This type of event is very common in the healthcare process for example, the process of patient transfer inside a hospital which starts when a nurse creates a call for transfer, the call might be updated or cancelled, then the call should be acknowledged and the outcome should be recorded.

The second category is ontological events which have a semantic relation with a main activity. For example, an admission activity can have a number of sub activities where the patient may have been admitted into different wards such as Medical Intensive Care Unit (MICU) or Coronary Care Unit (CCU). Our hypothesis is that mapping fine-grained activity into one main activity will simplify the patient pathway model and reduce activity numbers to help surface interesting patterns.

Using our data model, the categories of fine-grain activity are relatively limited for some tables. Transactional events are located in *Call, Input* and *Prescription* tables while ontological events are located in *Admissions* and *Transfer* tables. Hence, mapping the fine-grain activity into main activity was done manually using the *Add Mapping of Activity Names* log enhancement filter in ProM. The activities are mapped as illustrated in Table 4.

Table 4. Mapping transactional and ontological activities

Transactional activity	Mapped activity	Ontological activity	Mapped activity
Call create	Call	Admit CCU	Admit
Call update	Call	Admit CSRU	Admit
Call acknowledge	Call	Admit MICU	Admit
Call outcome	Call	Admit SICU	Admit
Call first reservation	Call	Admit TSICU	Admit
Call current reservation	Call	Transfer CCU	Transfer
Input start	Input	Transfer MICU	Transfer
Input store	Input	Transfer CSRU	Transfer
Input comment	Input	Transfer SICU	Transfer
Input end	Input	Transfer TSICU	Transfer
Prescription start	Prescription		
Prescription end	Prescription		

Results
The results of this experiment shows that the number of different types of activities was reduced by nearly half of the previous processing step. Also, the number of events was reduced and consequently the mean of events per case is reduced.

On the other hand, the number of variations remained high and was not affected by mapping fine-grain activity. Table 5 shows some statistics of pathway characteristics after applying step 2 of event log processing.

We believe that the resulting event log from step 2 can be used as a baseline event log for applying an interval-based event selection method.

Table 5. Summary of the extracted diabetes event log after processing step 2

Pathway characteristics	
Admissions (cases)	296
Patients	264
Variations	100%
Activity	15
Events	210,139
Mean event per case	710
Minimum event per case	21
Maximum event per case	9246

5 The Rationale for an Interval-Based Event Selection Method

In this paper, we define outlier events based on the time interval between events. Our starting assumption is that an event is regarded as an outlier if it occurs more frequently than a threshold interval determined from the central tendency and measure of dispersion of intervals for that event.

We take into consideration that process mining focuses on capturing events that comply with the mainstream process. For instance, in the case of blood measurements, two successive measurements that occur within a short interval may occur because of an error in the measurement value. Therefore, removing one of those events will not lead to information loss as both events correspond to the same observation. This assumption is supported by some data observation as shown in Table 6.

Table 6. Example of observations from input activity

hadm-id	Time	Item-id	Amount	cgid	Status	Cancel reason
101659	**2137-02-27 23:00:00**	**221749**	**1.400105**	14953	Changed	0
101659	2137-02-27 23:00:00	225158	5.833345	14953	Changed	0
101659	**2137-02-27 23:35:00**	**221749**	**5.603825**	14953	Changed	0
101659	2137-02-27 00:45:00	225158	23.34927	14953	Changed	0
101659	2137-02-27 00:45:00	221749	6.970018	14953	Changed	0

The table shows events extracted from *Input* table. The first and third highlighted rows belong to the same observed item where item id = **221749** for the same patient and the same ICU number. Assuming the interval pattern of input activity is 1 h, the third row displays that this event occurred after 35 min from the previous one. It appears that this event is repeated because the care giver has changed the amount of the intake item.

5.1   Interval-Based Event Selection Method

In this section, some formal definitions are provided to avoid any ambiguity in the method. The definitions are illustrated in Fig. 3.

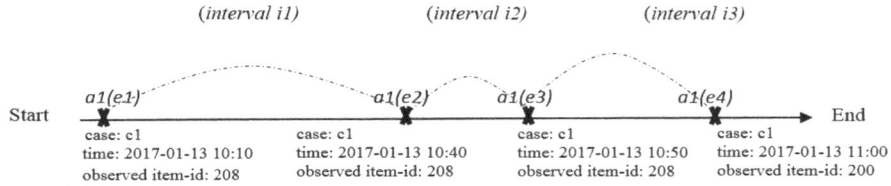

Fig. 3. Example of interval-based selection method definitions

Definition 1 (case c): is a single episode of care consists of different activities.
Definition 2 (activity a): An activity is an event class.
Definition 3 (event e): $e \in a$ where, e is an instance of a and has a timestamp and other attributes.
Definition 4 (observed item x): is a distinguished attribute of an event e.
Definition 5 (consecutive events (e1, e2)): $e1$, $e2$ are consecutive events $\in$ same activity a.
Definition 6 (interval i): is the period of time between *consecutive events (e1, e2)*.

Our approach has several steps:

1. Create histograms of intervals i for each activity.
2. Use histograms to determine the central tendency and dispersion of the intervals to calculate a threshold value to identify outliers. Examples may be the mean, median and standard deviation depending on the shape of the distribution.
3. For each *case c* in the *log*, get *activity a* and compare the interval between its *consecutive events (e1, e2)* until the end of the *case*. The interval between each *consecutive events (e1, e2)* is computed by finding the time difference between *e1* and *e2*.
4. If the interval between *consecutive events (e1,e2)* is less than the threshold value of that activity and both events occurred on the same observed item then, remove the second event as this event is an outlier based on our assumption. Otherwise, keep the second event because it belongs to a different item.
5. If the interval is equal or longer than the threshold value, keep both events because they comply with the pattern.

5.2   Interval-Based Event Selection Method and the MIMIC-III Database

The proposed approach is evaluated on our data from MIMIC-III database. Histograms are used to illustrate the interval between events for all activities. Figure 4 shows interval histograms for some activities such as *Lab* and *Notes* activity. The threshold

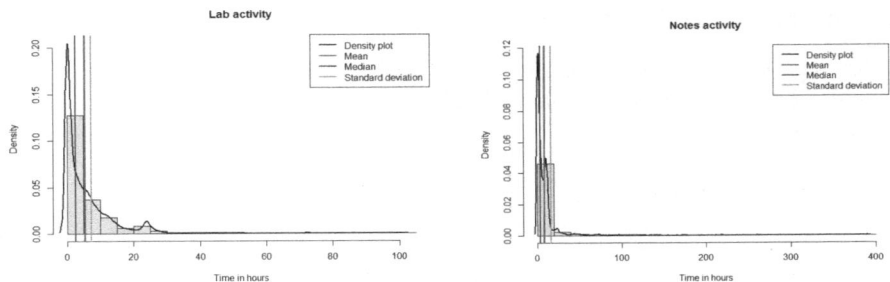

Fig. 4. Interval histogram for Lab and Notes activity

value is selected based on the mean for most of the activities because it represents the majority of the cases however, it depends on the interval distribution and the user preferences.

Table 7 shows the threshold interval values for repeated activities in MIMIC-III. These values are used for filtering the events for next experiments.

Table 7. The selected threshold interval value of repeated activities

Activity	Interval length	Activity	Interval length
Call	1.5 h	Chartevent	34.6 min
Cpt event	27.5 h	Labevent	6.0 h
Prescription	25.2 h	Input	1.1 h
Microbiology	66 h	Noteevent	8.9 h
Output	1.6 h	Transfer	52.8 h

For instance, In order to filter the Chartevent activity with interval value of 34.6 min. Our method aims to eliminate outlier events which have occurred in a time that is shorter than the selected threshold and belong to the same observed item for example, blood pressure. Let x, y, z be consecutive events of the Chartevent activity which occurred at the times 03:54, 04:00, 04:30 respectively for the patient ID 100908 as shown in Fig. 5.

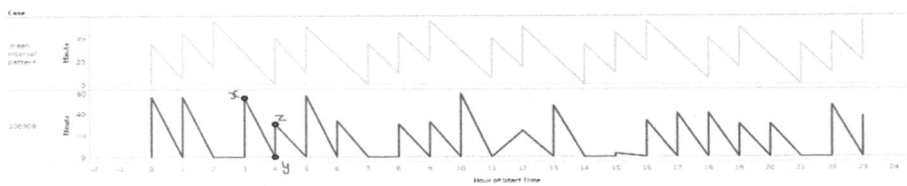

Fig. 5. Example of remove an outlier event from Chartevent activity

The interval between x and y is computed which is 6 min hence, this is shorter than the threshold value. Then, x and y are checked if they are events for the same *observed item*, both for measuring blood pressure. If x and y belong to the same item this means the same item is checked twice therefore, y event is removed as it is an outlier.

After removing y, the interval between x and z is computed as they became consecutive events. The interval = 36 min which is longer than the threshold value. Hence, keep the event z and move forward to compare it with the next event.

5.3   Results and Evaluation

Removing the outliers events from the event log using interval based event selection has reduced the number of events, mean event per case and maximum events per case while other pathway characteristics such as variations, number of activity and the minimum number of events have not affected. The following Table 8 shows some statistics of pathway characteristics after applying this approach.

Table 8. Summary of the diabetes event log after interval based event selection

Pathway characteristics	
Admissions (cases)	296
Patients	264
Variations	100%
Activity	15
Events	208580
Mean event per case	705
Minimum event per case	21
Maximum event per case	9189

Moreover, this method has different impact on the activities. Some activities have been affected strongly by removing the outliers' events such as *Prescription* and *Cptevent* where 331 outliers' events are removed from *Prescription* and 248 in *Cptevent* activity. On the other hand, *Chartevent* and *Output* activities have the least impact with 56 outliers' events in *Chartevent* and 64 in *Output* activity.

We have evaluated our approach by comparing it with the existing log preparation techniques in ProM that tried to remove the outliers of repeated activities. Despite the significant reduction of events number using *merge subsequent event* filter as shows in Table 9, there are a number of limitations as discussed earlier. The first and second options of merging ignore the time aspect between events and concentrating on reducing the number of events without consideration of losing time information. The third option of merging may result in misleading event duration.

To sum up, our simple interval based event selection method can outperform the current approach of repeated activity filtering. This is because an interval based approach takes into account temporal perspective between events unlike current technique that merge events regardless of the time perspective between them.

Table 9. Pathway characteristics using merge subsequent event

Pathway characteristics using merge subsequent event plugin	
Admissions (cases)	296
Patients	264
Variations	100%
Activity	15
Events	133887
Mean event per case	452
Minimum event per case	15
Maximum event per case	5229

The Impact of Interval-Based Selection on Model Fitness and Precision

Some performance measurements in the ProM framework such as fitness and precision have been used to evaluate our approach.

We have used the inductive miner because it is a robust, it generates sound model and has reliable precision and fitness measurements. It is used to generate a Petri net for both original and cleaned event log. Two variants of inductive miner are used, inductive miner (IM) and inductive miner–infrequent (IMi). The IM tries to divide the log into sub-logs by finding the best cut points between traces. It guarantees the rediscover-ability for all traces. The IMi is a variant of IM with focusing on generating more precise model by discarding the infrequent traces among all divided sub-logs.

Model precision measurement in ProM can be calculated using different formulas. Generally, the precision estimates how many traces can be generated from the model which are not observed in the event log. In ProM, a Petri net model should be built on the log. In our investigation we used Inductive Miner to generate the Petri net and use this model and the logs for measuring precision.

We used the alignment based precision method [17] because this method is more reliable as it does not penalize the model for allowed deviations of the traces that are not observed in the log however, it penalizes the model on the traces that are extremely dissimilar of the observed traces using alignment score to identify traces similarity. We found that, our method improved model precision without reducing model fitness as shown in Table 10.

Table 10. Precision and fitness comparison between original and cleaned log

Process miner	IM		IMi	
	Fitness	Precision	Fitness	Precision
Original event log	1	0.14	0.95	0.25
Cleaned event log	1	0.30	0.95	0.44

The Impact of Interval-Based Selection on Event Log Visualization

Event log visualization is a significant tool for exploring the data. We have visualized the baseline event log beside the cleaned event log to test the effect of our method on activity pattern detection. Figures 6 and 7 display a dotted chart visualization using ProM for *prescription* activity for some diabetes patients. The Y axis represents several patients' admissions and X axis represents time since case starts. We can see the activity pattern is clear after removing outliers in Fig. 7. In contrast, the activity pattern in Fig. 6 cannot be captured easily where some intervals between *prescription* events are very short while others are consistent with the mean interval.

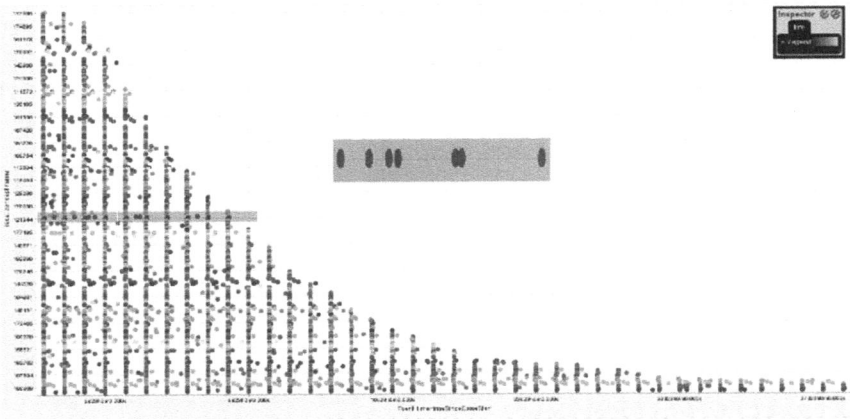

Fig. 6. Dotted chart of baseline event

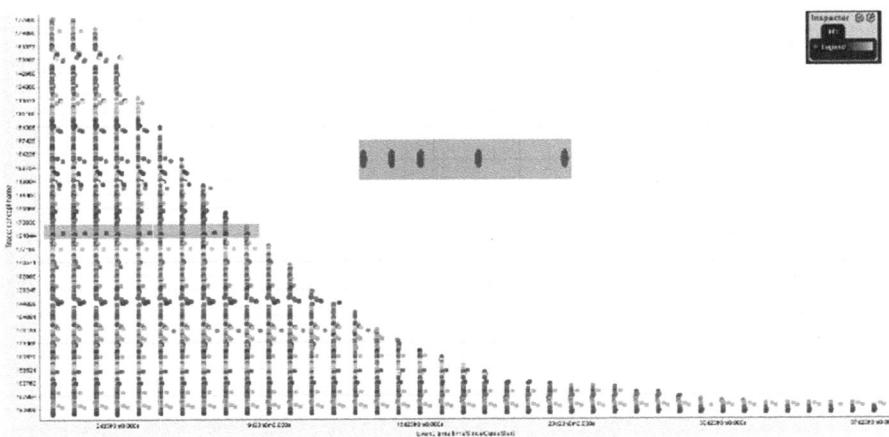

Fig. 7. Dotted chart of cleaned event

6   Conclusion and Future Work

Interval-based event selection method is a technique that can be used for event log preparation as a preliminary step before applying process model discovery techniques. Our novel approach aimed to reduce the variations by filtering outlier events based on the mean interval of activities but the median or a number of standard deviations from the mean could also be used. The method improved model quality without reducing model fitness and has the potential of improving pattern visualization. In some situations, using the mean interval to identify deviation threshold may not be the best choice because it depends on the data distribution.

Furthermore, activity duration has an influence on the interval pattern hence, using this approach on a reliable activity duration will produce better results. This method is tested using a Petri net generated by Inductive miner which is not an advanced miner that can deal with repeated activity as the case of Genetic miner. We believe that, using our approach with an advanced miner will improve model quality. Future research aims to integrate the existence techniques of extracting N-gram patterns with the interval-based cleaning method where an interval of a pattern rather single event type will be used as filtering threshold. Further work is needed to address the evaluation limitations of the interval-based approach that is presented in this paper.

References

1. Van der Aalst, W.: Process Mining: Discovery, Conformance and Enhancement of Business Processes. Springer, Berlin (2011)
2. Van Dongen, B.F., Medeiros, A.K.A., Verbeek, H.M.W., Weijters, A.J.M.M., Aalst, W.M. P.: The ProM framework: a new era in process mining tool support. In: Ciardo, G., Darondeau, P. (eds.) ICATPN 2005. LNCS, vol. 3536, pp. 444–454. Springer, Heidelberg (2005). doi:10.1007/11494744_25
3. Mans, R.S., Van der Aalst, W.M.P., Vanwersch, R.J.B.: Process Mining in Healthcare: Evaluating and Exploiting Operational Healthcare Processes. Springer, Heidelberg (2015)
4. Weiskopf, N.G., Weng, C.: Methods and dimensions of electronic health record data quality assessment: enabling reuse for clinical research. J. Am. Med. Inform. Assoc. **20**(1), 144–151 (2013)
5. Van der Aalst, W., Adriansyah, A., et al.: Process mining manifesto. In: Daniel, F., Barkaoui, K., Dustdar, S. (eds.) BPM 2011. LNBIP, vol. 99, pp. 169–194. Springer, Heidelberg (2012). doi:10.1007/978-3-642-28108-2_19
6. Bose, R.J.C., Mans, R.S., Van der Aalst, W.: Wanna improve process mining results? In: 2013 IEEE Symposium on Computational Intelligence and Data Mining (CIDM). IEEE (2013)
7. de San Pedro, J., Cortadella, J.: Discovering duplicate tasks in transition systems for the simplification of process models. In: La Rosa, M., Loos, P., Pastor, O. (eds.) BPM 2016. LNCS, vol. 9850, pp. 108–124. Springer, Cham (2016). doi:10.1007/978-3-319-45348-4_7
8. Vázquez-Barreiros, B., Mucientes, M., Lama, M.: Mining duplicate tasks from discovered processes. In: ATAED@PetriNets/ACSD (2015)
9. Van der Aalst, W., et al.: Process mining: a two-step approach to balance between underfitting and overfitting. Softw. Syst. Model. **9**(1), 87 (2010)

10. Broucke, S.V.: Advances in process mining: artificial negative events and other techniques (2014)
11. da Silva, L.F.N.: Process mining: application to a case study (2014)
12. Lu, X., Fahland, D., Biggelaar, F.J.H.M., Aalst, W.M.P.: Handling duplicated tasks in process discovery by refining event labels. In: La Rosa, M., Loos, P., Pastor, O. (eds.) BPM 2016. LNCS, vol. 9850, pp. 90–107. Springer, Cham (2016). doi:10.1007/978-3-319-45348-4_6
13. Suriadi, S., et al.: Event log imperfection patterns for process mining: towards a systematic approach to cleaning event logs. Inf. Syst. **64**, 132–150 (2017)
14. Johnson, A.E., et al.: MIMIC-III, a freely accessible critical care database. Sci Data **3**, 160035 (2016)
15. MIMIC medical database. MIMIC-III critical care database (2015). https://mimic.physionet.org/gettingstarted/access/. Accessed 9 Mar 2017
16. Kurniati, A., et al.: The assessment of data quality issues for process mining in healthcare using MIMIC-III, a publicly available e-health record database (2017)
17. Adriansyah, A., et al.: Measuring precision of modeled behavior. IseB **13**(1), 37–67 (2015)

Soundness of Decision-Aware Business Processes

Kimon Batoulis$^{(\boxtimes)}$ and Mathias Weske

Hasso Plattner Institute, University of Potsdam, Potsdam, Germany
{Kimon.Batoulis,Mathias.Weske}@hpi.de

Abstract. With the recent release of the Decision Model and Notation (DMN) specification, standardized decision models can be designed to represent the decisions required for executing business processes. Outsourcing decision logic from process to decision models leads to a separation of concerns and therefore to decision-aware business processes. However, no exhaustive considerations regarding the soundness of the integration of the two types of models have been made so far. Classical soundness checking only looks at the control-flow of a process model. In this paper, we formally define soundness criteria for decision-aware processes that ensure that the process can continue after a decision has been taken, and that all activities following the decision can be executed. A scalable implementation and an analysis of models from participants of an online course on process and decision modeling as well as a from a BPM project of a large insurance company demonstrate the benefits of our contribution.

Keywords: Process modeling · Decision modeling · BPMN · DMN · Soundness

1 Introduction

Business process models are valuable assets for business organizations to capture and improve their business processes. These models represent the logical ordering of work activities that need to be conducted in order to achieve a business goal [18]. In practice, processes that consist of just a single flow of activities are rare. Often, there are points during execution at which one of several alternative branches must be selected in order to continue. Hence, a decision must be made. The information required for such decisions as well as the logic to actually make them can be expressed in a decision model. Thus, the process will "call" this model with some input data and expects a return value—the outcome of the decision—that can then be used to select one of the alternative branches.

The industry standard to represent process models is the Business Process Model and Notation (BPMN) [13], which is complemented by a recently published standard for modeling decisions: the Decision Model and Notation (DMN) [14]. These two standards are designed to be used in conjunction which leads to the notion of *decision-aware* business processes [17]. A decision-aware business process understands that there is a substantial difference between tasks

© Springer International Publishing AG 2017
J. Carmona et al. (Eds.): BPM Forum 2017, LNBIP 297, pp. 106–124, 2017.
DOI: 10.1007/978-3-319-65015-9_7

that perform work and tasks that make decisions based on data and logic, and that the details of the latter should be outsourced to decision models. Yet, so far, no thorough considerations have been made regarding the soundness of decision-aware business processes. Some preliminary research has been conducted by us [19]. In this work, we semi-formally describe criteria that must be fulfilled in order to ensure consistency between process and decision models. For example, each outcome of the decision model must be selected by at least one of the alternative branches in the process model. Analogously, for each branch in the process model, there must be an outcome of the decision model such that this branch is selected.

In this paper, we first argue that the traditional soundness criteria defined for process models [2] yield incorrect results when being applied to process models that are used in conjunction with decision models because they only consider control flow information. Based on this argumentation, we define, implement and empirically evaluate the notion of decision-aware soundness for processes together with corresponding soundness criteria. These criteria can be applied to all possible types of DMN decision models, but particular emphasis is put on the standardized concept of DMN decision tables, allowing overlapping rules and any number of output variables.

The remainder is structured as follows. We motivate our approach by introducing a running example and by discussing limitations of existing methods in Sect. 2. Section 3 lays the foundations for our work, so that in Sect. 4 soundness for decision-aware business processes can be defined. A corresponding implementation as well as an empirical evaluation are described in Sect. 5. Section 6 is devoted to related work, and Sect. 7 concludes the paper and discusses directions for future research.

2   Motivation and Running Example

Our approach is motivated by and explained along an example process we derived from booking train tickets with the *Deutsche Bahn* (*DB*), a German railway company[1]. While entering travel details to book a ticket, it is possible to indicate that one is eligible for a discount. For example, *DB* sells discount cards, called *BahnCard*, that are valid for one year. These cards are available in two types that can be used when booking a ticket: *BahnCard 25* and *BahnCard 50*, which grant card holders to get 25% and 50% discount respectively on ticket prices. Furthermore, the card is associated with a bonus system, called *bahn.bonus*. Whenever a ticket is purchased, the card holder receives one point for every Euro of the ticket price. Collected points can be redeemed for goods and services. Let us assume for this example that given a certain amount of points, the ticket prices can be discounted even further.

Since *DB* is eager to increase customer loyalty, it offers customers that do not indicate ownership of such a card

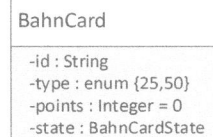

Fig. 1.   *BahnCard* data model

---

[1] http://www.bahn.de/p_en/view/index.shtml.

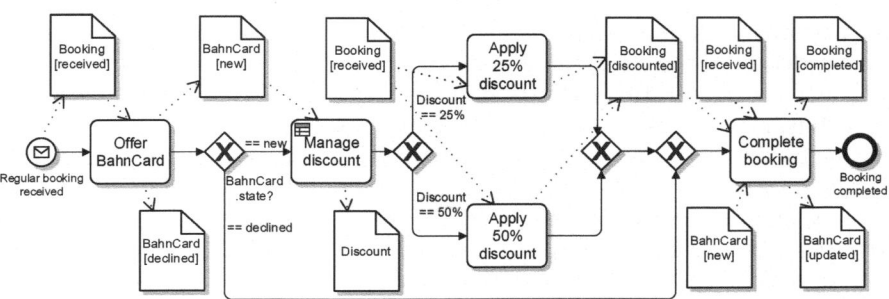

Fig. 2. Process model for bookings that do not indicate ownership of a *BahnCard*

while booking a ticket to buy it together with the ticket in order to get a discount. The corresponding process that we derived is illustrated in Fig. 2. After a regular booking—one that does not indicate eligibility for a discount—is received, a *BahnCard* is offered to the customer. If she decides not to buy it, the booking is completed with the regular price. In the other case, a new *BahnCard* object is created for the customer that holds information about its type (25 or 50) and the number of points, which after creation is 0. The *BahnCard* data model is illustrated in Fig. 1. Next, the amount of the discount that the customer is eligible for is determined. This is done by a business rule task that is linked to the decision model represented in Fig. 3. The data object *BahnCard* that is read by *Manage discount* in the process model is equivalent to the input data in the decision model. After the decision model has been evaluated and the result returned to the process, *Manage discount* will write it into the *Discount* data object. The value of that object is then evaluated at the split gateway and the corresponding path is taken to apply a discount to the booking. The last activity completes the booking and possibly adds bonus points to the customer's card, given that she chose to buy one.

One might now want to check the process model for soundness, as defined in [2]. Clearly, the process is structurally sound because it contains exactly one start and one end event and every node is on a path from the start to the end event. Furthermore, from a behavioral point of view, (i) regardless of what the process does, each case that starts in the initial state will eventually reach the final state; (ii) when the final state is reached, nothing happens anymore in the process; (iii) all activities can participate in at least one execution. Therefore, one might conclude that the process is sound.

However, notice that the set of outputs that are possible in the decision table in Fig. 3 (25%, 50%, 60%, 70%) is larger than what the process model in Fig. 2 actually expects (25% and 50%). Consequently, the soundness check just described produces an incorrect result because criterion (i) is actually violated since the process cannot continue when $Discount \in \{60\%, 70\%\}$.

Motivated by this example, we formulate soundness criteria that can be applied to identify such situations. For that purpose, in Sect. 4, we first describe

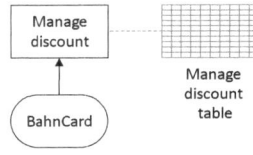

UC	BahnCard.type (25, 50)	BahnCard.points Number	Discount (25%, 50%, 60%, 70%)
1	25	< 1000	25%
2	50	< 1000	50%
3	25	>= 1000	60%
4	50	>= 1000	70%

(a) Decision model linked to the *Manage discount* task

(b) Decision table *Manage discount* for determining a discount. This table is associated with the decision node in the model on the left.

Fig. 3. Decision model for *Manage discount*

how the set of outputs of any kind of DMN decision table can be derived. Based on that, we define the notion of soundness for decision-aware business processes along with corresponding verification criteria. These criteria will determine that the example above is not sound.

3 Foundations

DMN defines two levels for modeling decision models, the decision requirements level and the decision logic level. The first one represents how decisions depend on each other and what input data is available for the decisions. Therefore, these nodes are connected with each other through information requirement edges. A decision may additionally reference the decision logic level which describes the actual decision logic applied to take the decision. Decision logic can be represented in many ways, e.g., by an analytic model or a decision table. The consistency criteria defined in this paper are exemplified using decision tables because they are standardized in DMN.

In general, a decision table describes a relation between a set of input values and a set of output values. Each *input* has a *domain* over which logical expressions can specify *conditions*. Given a value for each input, if the conjunction of logical expressions of all inputs evaluates to true, the corresponding output values can be determined, taken from the domains of the outputs. All the possibilities of relating different inputs to outputs are represented in a tabular manner such that each row of the table corresponds to a *rule*. It is possible to define overlapping rules, i.e., rules that match for the same set of input values. These rules do not necessarily need to map to the same output values. Thus, a decision table in itself is not guaranteed to define a right-unique relation (i.e., a function). To avoid these kinds of ambiguities, DMN defines several *hit policies* that determine a unique output in various ways in case more than one rule is matched. These policies can be divided into two classes: the policies *unique, any, priority, first* ensure that only one rule's output is returned. *Output order* and *rule order* return a list (or sequence) of outputs, namely the outputs of each matching rule, in a particular order.[2] More detailed information about these policies is provided in Sect. 4 and in the DMN standard [14].

---

[2] We omit the *collect* policy because it is ambiguous with respect to this classification.

Additionally, a *completeness indicator* specifies whether or not the relation is left-total, i.e., if all possible combinations of input values are covered by the decision table. Therefore, a DMN decision table specifies either a function or a partial function. Decision tables can then be defined formally as follows (adapted from [16]):

Definition 1 (DMN decision table). A decision table dt is a tuple $(I, O, dom, R, prio, p, c)$ where:

- $I = \{i_1, i_2, \ldots, i_n\}$ is a set of $n > 0$ input variables.
- $\forall i \in I, dom(i)$ is the domain of input variable i, where dom is a function mapping a variable to its domain.
- $O = \{o_1, o_2, \ldots, o_m\}$ is a set of $m > 0$ output variables.
- $\forall o \in O, dom(o)$ is the domain of output variable o.
- Given the input value combinations $IV = \prod_{j=1}^{n} dom(i_j)$ and the set of output value combinations $OV = \prod_{j=1}^{m} dom(o_j)$, the *finite* set of $q > 0$ rules is defined as $R = \{r_1, r_2, \ldots, r_q\}$, where $|R| = q, q \in \mathbb{N}$ and $\forall r \in R : r \subseteq IV \times OV$.
- $prio : R \rightarrow \{1, \ldots, |R|\}$ assigns each rule a priority visualized by its rule number in the leftmost column of the table. If no priority is explicitly given, it is implicitly given by the graphical ordering of the rules in the table.
- $p : dt \rightarrow \{unique, any, priority, first, output\ order, rule\ order\}$ is a function that assigns each decision table a hit policy.
- $c : dt \rightarrow Bool$ is the completeness indicator, i.e., a function that indicates whether or not a decision table is complete.

Example. Consider the decision table *Manage discount* depicted in Fig. 4. This table can be formally defined as follows.

UC	BahnCard.type *(25, 50)*	BahnCard.points *Number*	Discount *(25%, 50%, 60%, 70%)*
1	25	< 1000	25%
2	50	< 1000	50%
3	--	>= 1000	60%

Fig. 4. Decision table *Manage discount* for determining a discount

- $I = \{BahnCard.type, BahnCard.points\}$, $O = \{Discount\}$
- $dom(BahnCard.type) = \{25, 50\}$, $dom(BahnCard.points) = \mathbb{N}$, $dom(Discount) = \{25\%, 50\%, 60\%, 70\%\}$.
- $R = \{r_1, r_2, r_3\}$, where $\forall r \in R : r \subseteq (\{25, 50\} \times \mathbb{N}) \times \{25\%, 50\%, 60\%, 70\%\}$. For example, $r_3 = (\{25, 50\} \times \{n \mid n \in \mathbb{N} \wedge n \geq 1000\}) \times \{60\%\}$, because the expression "--" allows any value for the variable *BahnCard.type*.
- $prio = \{(r_1, 1), (r_2, 2), (r_3, 3)\}$.
- $p(Manage\ discount) = unique$ (denoted by U in the upper left corner).
- $c(Manage\ discount) = true$ (denoted by C in the upper left corner).

Decision models made up of a decision requirements graph and decision logic based on decision tables can then be defined as follows.

Definition 2 (Decision model). A decision model dm is a tuple (D, InD, IR, Out, tab) where:

- D is the set of decision nodes.
- $d_{top} \in D$ is the top-level decision.
- InD is the set of input data nodes.
- $IR \subseteq \{D, InD\} \times D$ is the set of directed information requirement edges.
- $tab : D \to DT$ assigns each decision $d \in D$ a decision table.
- Out is the *finite* set of possible outputs of the decision model, which is equal to the set of possible outputs of its top-level decision d_{top}. Clearly, a decision model might have sub-decisions, which also have outputs. These outputs, however, are not part of Out since they are only used internally by the decision model.

Figure 3 shows a simple decision model consisting of one decision associated with a decision table and connected to an input data node. Lastly, process models, such as the one depicted in Fig. 2, and fragments thereof are defined in the following way.

Definition 3 (Process model). A process model m is a tuple $(N, DO, C, CF, DF, \alpha, \xi, \delta)$ where:

- $N = A \cup G$ is a finite non-empty set of control flow nodes, with set A of activities and set G of gateways.
- $A_D \subseteq A$ is the set of business rule tasks.
- DO is a finite set of data object nodes.
- C is a finite set of conditions. Each condition $cond_i \in C$ is associated with one or more variables over which it specifies unary tests.
- $CF \subseteq N \times N$ is the control flow relation such that each edge connects two control flow nodes.
- $DF \subseteq (DO \times A) \cup (A \times DO)$ is the data flow relation indicating read respectively write operations of an activity with respect to a data node.
- Let Z be a set of control flow constructs. Function $\alpha : G \to Z$ assigns to each gateway a type in terms of a control flow construct.
- $\forall g \in G : \alpha(g) \in \{exclusive, inclusive, complex\} : \xi : (G \times N) \cap CF \nrightarrow C$ is a function that assigns conditions to control flow edges originating from gateways representing a decision point.
- $\delta : A_D \to DM$ is a function assigning a decision model to each business rule task.

Definition 4 (Process fragment). Let m be a process model. A *process fragment* $f = (N', DO', C', CF'DF', \sigma, \gamma, \eta)$ is a connected subgraph of process model m.

When linking DMN models to BPMN business rule tasks that affect the control flow of a business process, the typical fragment that is encountered is that of a split gateway with two or more outgoing edges (or branches) where the branch conditions correspond to the outputs of the decision model. These *decision fragments* are defined as follows.

Definition 5 (Decision fragment). Let f be a process fragment of a process model. Fragment f represents a decision fragment if it starts with exactly one business rule task t, that is directly followed by a split gateway g, where $\alpha(g) \in \{exclusive, inclusive, complex\}$. g is directly followed by at least two activities. t reads and writes at least one data object respectively, and each outgoing branch of g is annotated with a condition.

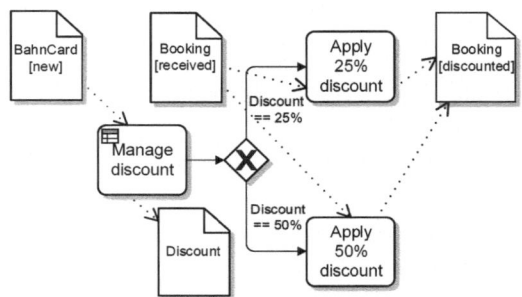

Fig. 5. Decision fragment for managing a discount

Example. Consider the process fragment in Fig. 5. This fragment represents a decision fragment with business rule task *Manage discount*, an exclusive split gateway and two activities that act on the result of the decision. The decision is made based on the data in the *BahnCard* data object, the result is written into the *Discount* data object and its value is tested at the outgoing branches of the gateway. Accordingly, one of the subsequent activities is applied to a *Booking* data object.

4   Defining Soundness for Decision-Aware Business Processes

In the following, we define the notion of soundness for decision-aware business processes. This notion has a simple structural and two more elaborate behavioral aspects, called *decision deadlock freedom* and *dead branch absence*. We first informally describe structural requirements in Definition 6 and then formally define two verification criteria in Sects. 4.2 and 4.3 regarding the two behavioral aspects. For these definitions, an algorithm to compute the set of possible outputs of a decision table is required, described in Sect. 4.1. With the help of these criteria we can give a definition of decision-aware soundness in Sect. 4.4. The verification criteria build upon preliminary work described in [19] and are designed for situations in which the set of possible outcomes of the decision model are known at design time and the outcome of the decision model is processed by a decision fragment (cf. Definition 5). The criteria are thus checked locally for every decision fragment of the process model. Since decision tables are standardized in DMN, we use them for demonstration purposes. This is further discussed in Sect. 4.4.

Regarding structural consistency, the use of BPMN data objects and DMN input data must be consistent regarding the process and decision models under investigation. This is captured by the following definition:

Definition 6 (Structural Consistency). For each input data in the decision model there is a data object in the business process that is read by the business rule task. Moreover, the DMN input data and BPMN data objects reference the same data model or they are equal with respect to their contained attributes and information.

4.1   Determining the Possible Outputs of a DMN Decision Table

The fundamental idea of the behavioral criteria described in Sects. 4.2 and 4.3 is to compare the outputs of the decision model (i.e., the outputs of its top-level decision table) with the conditions of the branches of the decision fragment. For example, one needs to make sure that each output of the top-level decision table is covered by some branch condition. For simple, unambiguous decision tables, knowing the rules and their outputs may be enough to determine the set of outputs of the table. However, DMN defines hit policies that determine what happens if an input combination matches more than one rule. This is the case when those rules are *overlapping*. Two or more rules overlap if there exist input value combinations that match these rules at the same time. In these cases, determining the set of outputs of the table is not straightforward.

An example is given by the decision table in Fig. 6, henceforth called t. Rules 1, 2 and 3 of t overlap: they all match when $input1 = 1$ and $input2 = true$. In these cases—since t's hit policy is *rule order* (cf. letter R in the upper left corner)—the final table output will be a *sequence* of the outputs of rules 1, 2 and 3 (in that order): (x, y, z). Furthermore, rules 2 and 3 can additionally match

RC	input1 Number	input2 Boolean	output {w,x,y,z}
1	1	true	x
2	>= 1	true	y
3	--	true	z
4	--	false	w

Fig. 6. A decision table with overlapping rules (1, 2 and 3). This table's output can be more than the output of an individual rule.

without rule 1, if $input1 > 1$ and $input2 = true$. Therefore, the output set of t is given as follows: $output(t) = \{(x, y, z), (y, z), z, w\}$, because only rules 3 and 4 can be triggered individually. But whenever rule 1 is triggered, rules 2 and 3 are also triggered, and whenever rule 2 is triggered, rule 3 is also triggered. In the following, we will describe a method for determining the set of outputs $output(dt)$ of any kind of DMN decision table.

Generally, $output(dt) \subseteq \{out \mid out \in OV \vee out = (ov_j)_{j=1,...,l}, ov_j \in OV\}$ is the output set of a decision table dt described by R and p, i.e., the set of outputs it can produce given the rules and the hit policy. An element of $ouput(dt)$ can be the output of a single rule (i.e., $out \in OV$), or a sequence of outputs of several rules, in a particular order (i.e., $out = (ov_j)_{j=1,...,l}, ov_j \in OV$). However, in order to compare the set of outputs of a decision table with the branch conditions of a process model, we need to know *exactly* which outputs dt can produce. This depends on the hit policy of the table, such that for each policy we will describe how it is handled.

If the table has a *unique* policy, there are no overlapping rules, such that each rule is matched individually. Therefore, the output of each rule is added to the output set $output(dt)$ of the decision table. Of course, there may be rules having the same output. Yet, since $output(dt)$ is defined to be a set, each unique output will only be added once. For this reason, the *any* policy can be treated the same way. This policy allows overlapping rules but requires those rules to have the same output, such that it does not matter which rule's output is chosen.

The situation is different for the other hit policies. In these cases, the table may have overlapping rules with different outputs, and depending on the hit policy either one (single-hit) or all of the matching rules (multi-hit) are triggered. So, what we are looking for are the sets of rules that can be matched by the allowed inputs to the table and from this inferring the sets of outputs that these sets of rules can produce based on the hit policy. Note that if two rules r_1 and r_2 are overlapping, it may be the case that the inputs for which r_1 matches are a subset of those for which r_2 matches. For example, in the table in Fig. 6, rule 2 is always matched together with rule 3. Therefore, an overlapping rule is matched together with the rules it is overlapping with and is optionally also matched individually (cf. rule 3 in the table above).

To determine the sets of rules matched by the allowed inputs to the table, we extended an algorithm for finding overlapping rules in a DMN table described in [4]. This algorithm computes maximal sets of overlapping rules for a given DMN table. We enhanced this algorithm to not only report *maximal* sets of rules that can be matched by an input, but to report *any subset* of the set of rules of the table that can be matched separately by some input. For example, the algorithm in [4], will only report that rules 1, 2 and 3 of the table t above are overlapping. Our extension of this algorithm will additionally determine that rules 1, 2 and 3 match together (for $input1 = 1$, $input2 = true$); rules 2 and 3 match together (for $input1 > 1$, $input2 = true$); rule 3 matches individually (for $input1 < 1$, $input2 = true$); and rule 4 matches individually (for $input2 = false$). For information about the steps that we added to the algorithm in [4], we refer the reader to our implementation website: https://bpt.hpi.uni-potsdam.de/Public/BpmnDmnSoundness/Algorithm.

Let *matchingSet* be the set that we just computed, containing sets of rules that can be matched by inputs to the table. For example, $matchingSet(t) = \{\{1, 2, 3\}, \{2, 3\}, \{3\}, \{4\}\}$. For each *match* $\in$ *matchingSet* that contains only one rule, this rule's output can just be added to *output* because there are no ambiguities. So, the outputs of rule 3 and 4 of table t, namely z and w, can already be added to $output(t)$. If, however, *match* contains more than one rule, it depends on the hit policy which of these rules can actually be *triggered* and in which order. The hit policies *priority* and *first* are single-hit policies, that return the output of a single rule only. Therefore, to select the rule whose output shall be returned, the function *prio* (cf. Definition 1) should be applied yielding the priority of a rule. In case of the *priority* policy, the rules are explicitly assigned a priority. For *first* policies, the *prio* function is defined implicitly by the ordering of the rules in the table. Hence, the rules in *match* are sorted according to their priorities in descending order and the first rule is added to *output*. The multi-hit policies *output order* and *rule order* behave analogous to *priority* and *first*, the only difference being that after the rules have been sorted according to the *prio* function, all rules are returned in that order. Thus, in our example, since t's hit policy is *rule order*, the outputs (x, y, z) (stemming from rules 1, 2 and 3 matching together) and (y, z) (from rules 2 and 3) are added to $output(t)$, thereby completing the set of possible outputs of t.

This concludes the description of our method for determining the set of outputs $output(dt)$ of any kind of DMN decision table. The behavioral criteria that are described in the following sections are based on a decision fragment f containing business rule task a_D, gateway g, and edge conditions C. Furthermore, $dm = \delta(a_D)$ is the decision model associated with a_D. It contains a set of decisions D and a corresponding set of decision tables DT. $dt_{top} = tab(d_{top})$ is the decision table of the top-level decision d_{top}. Each condition $cond_i \in C$ specifies a unary test that can be applied to the output values $out \in output(dt_{top})$ of the top-level decision table, i.e., $cond_i(out)$ yields *true* if out satisfies the condition $cond_i$; otherwise, it yields *false*.

4.2   Decision Deadlock Freedom

The *decision deadlock freedom* criterion makes sure that integrating a process model with a decision model does not lead to a deadlock. From a behavioral point of view, before *and* after the decision is executed a deadlock can occur. In the former case, even if process and decision model reference the same data object definition, the rules of the decision table reading the data object might not cover its entire domain. Stated differently, the data object (or one of its attributes) could assume a value for which there is no rule in the decision table. Consequently, the decision cannot be executed and the process is stuck. Therefore, each table of the decision model is required to be *complete*. Additionally, after the decision was made, we need to confirm that each possible output of the decision table is covered by at least one edge condition of the decision fragment it is connected to. If this is not the case, the decision may return a value for which there exists no alternative in the process. Hence, it will again deadlock and is not sound.

Let us first investigate the completeness property of a table in more detail. According to the DMN standard, a decision table is complete if "it produces a result for every possible case" [14]. Let PC be the set of possible cases. What are the elements of PC? Referring to Definition 1, one might say that $PC = IV$, the possible input value combinations of a decision table. In said definition, we stated that $IV = \prod_{j=1}^{n} dom(i_j)$, where $n = |I|$, the number of input variables. Therefore, IV contains all combinations of values from the domains of the input variables. What exactly are the domains of the input variables? From Definition 2 we know that a decision table that is part of a DMN decision model is associated with a decision node. Therefore, the input variables of a table are determined by the elements that the decision node is connected to via its information requirement edges. A connected node can either be an input data node or another decision node. In the former case, the domain of the input variable is simply given by the domain of the input data node. Regarding the latter case, from Sect. 4.1 we know that the set of possible outputs of a decision table is different from the domain of its output variable. For example, there might not be a rule for every possible value of the output variable's domain or several output values might be combined in a list based on a multi-hit policy. Therefore, in the formula $IV = \prod_{j=1}^{n} dom(i_j)$, if i_j comes from a sub-decision table, $dom(i_j)$ is equal to the output set of that table, rather than the domain(s) of its output variable(s).

This leads us to the definition of table completeness as required for decision deadlock freedom:

Definition 7 (Table completeness). A decision table is complete if and only if

$$\forall iv \in IV : \exists ov \in OV : \exists r \in R : iv \times ov \in r,$$

because for every possible case input value iv, there must be an output value ov, and at least one rule r of the table covers this input-output relation.

Let us now investigate the situation after the decision was made. In general, what does it mean that a decision table output is covered by an edge condition, or, stated differently, that an edge condition is satisfied by a table output? Imagine a table output such as the sequence (x, y). Consider the decision fragment in Fig. 7.

The upper branch's condition $output = (x, y)$ matches the entire sequence and $T1$ is selected as the next task. However, another possibility would be to have a condition that matches only a part of that sequence which is the case for the condition $x \in output$ of the lower branch. Both conditions *cover* the table output (x, y) since they evaluate to true when applied to that output. Although the lower branch does not consider the full sequence this does not lead to any problems regarding deadlock freedom—the process can still continue.

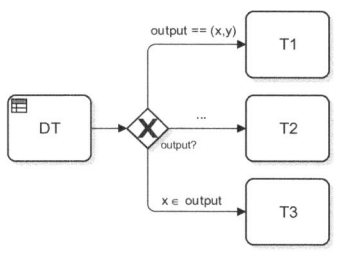
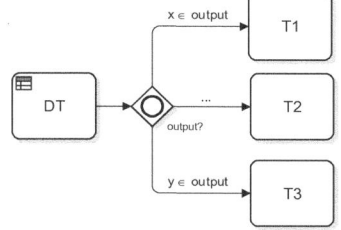

(a) Decision fragment with an exclusive gateway

(b) Decision fragment with an inclusive gateway

Fig. 7. Different possibilities of covering table outputs by edge conditions

Another example, is given in Fig. 7, where an inclusive gateway is used. In this case, several conditions matching only parts of a sequence can be combined to match the full sequence. This is because the inclusive gateway can activate more than one branch at a time. Thus, given the table output (x, y), this fragment will execute tasks $T1$ *and* $T3$. Therefore, both fragments in Fig. 7 cover the output (x, y) of the table in Fig. 6, but depending on the type of the gateway they can select a different number of branches.

Given the definitions and arguments above, we can formally define the *decision deadlock freedom* criterion as follows.

Definition 8 (Criterion: Decision deadlock freedom). The *decision deadlock freedom* criterion is satisfied if and only if

$$\forall dt \in DT : c(dt) = true \land \qquad \text{(table completeness)}$$
$$\forall out \in output(dt_{top}) : \exists cond_i \in C : cond_i(out) = true. \quad \text{(output coverage)}$$

Consider the process and decision models in Fig. 8. The criterion is violated for two reasons: first, the decision table is incomplete, because no rule considers the case when $BahnCard.points = 1000$. Second, the decision fragment does not cover the output 60% of the decision table. Note that also the value 70% could theoretically be an output of the decision table because it is in the domain of *Discount*. However, no rule exists that actually produces this output. Therefore, it would be sufficient for the decision fragment to additionally cover the output 60% since only the actual outputs of the decision table can occur.

4.3   Dead Branch Absence

The *dead branch absence* criterion makes sure that integrating a process model with a decision model does not lead to the process model having dead branches. More precisely, it verifies that each branch of the decision fragment is reachable through at least one output of the decision table, i.e., each condition of the decision

U	BahnCard.type *(25, 50)*	BahnCard.points *Number*	Discount *(25%, 50%, 60%, 70%)*
1	25	< 1000	25%
2	50	< 1000	50%
3	--	> 1000	60%

(a) Decision table *Manage discount* for determining a discount

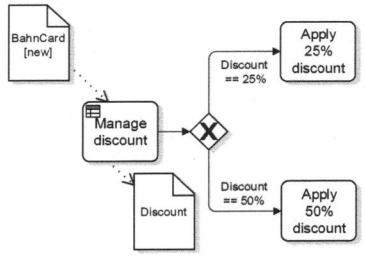

(b) Decision fragment that does not cover all decision outputs

Fig. 8. Example for a violation of the decision deadlock freedom criterion

fragment evaluates to true for at least one output of the table. If this is not the case, the process contains an activity that will never be executed during execution and is therefore not sound. The criterion is defined in the following way.

Definition 9 (Criterion: Dead branch absence). The *dead branch absence* criterion is satisfied if and only if

$$\forall cond_i \in C : \exists out \in output(dt_{top}) : cond_i(out) = true.$$

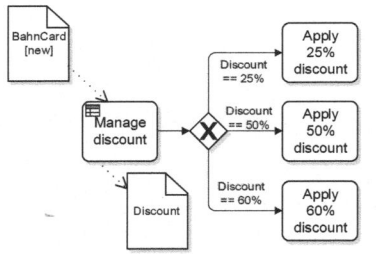

UC	BahnCard.type *(25, 50)*	BahnCard.points *Number*	Discount *(25%, 50%, 60%)*
1	25	< 1000	25%
2	50	< 1000	50%
3	--	>= 1000	50%

(b) Variant of the decision table *Manage discount* for determining a discount

Fig. 9. Example for a violation of the dead branch absence criterion

Consider the process and decision models in Fig. 9. The criterion is violated because the edge condition of the lowest branch of the decision fragment (== 60%) is not covered by the outputs of the decision table.

4.4   Decision-Aware Soundness

Given the criteria explained in this section, we can formulate soundness for decision-aware business processes as follows.

Definition 10 (Decision-aware soundness). A decision-aware BPMN process model is sound if and only if it is structurally consistent with all of its associated decision models and it is decision-deadlock free and contains no dead decision branches.

This soundness definition for decision-aware business processes is suitable for process and decision models for which analysis based on model level is possible. Naturally, this restricts the types of both process and decision models that can be analyzed with it.

Clearly, as the DMN standard proposes, the logic of each decision in a decision model can be expressed in any manner. Yet, our notion of soundness was formulated based on DMN decision tables. First of all, they are standardized in DMN and will presumably be used in the majority of decision models. But more importantly, they show the set of outputs that can be produced by them. This is different to a black box function of which we only know, if anything, the domain from which the set of outputs will be taken, i.e., the function's codomain. Therefore, our notion of soundness is not restricted to DMN decision tables only. Rather, it is suited for any kind of decision logic or function for which the set of values it maps to (its image) is known.

So far, we only associated decision models to process models containing decision fragments (cf. Definition 5). These decision fragments are characterized by the gateway following the business rule task to guide sequence flow. As is especially obvious from our running example, this is not the only way to process the outcome of a decision. Consider the process fragment in Fig. 10.

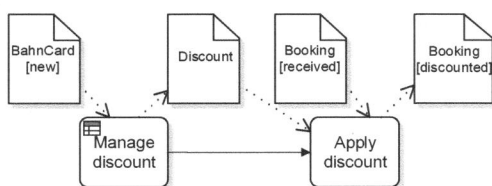

Fig. 10. Decision fragment that processes the decision outcome implicitly through reading a corresponding data object, not explicitly through a gateway

This fragment is very similar to the one shown in Fig. 5. The difference is there is only one activity processing the decision outcome that is now parameterized by the *Discount* data object written by the decision task. Consequently, it is not possible on model-level to check which values the process expects in order to continue. One would have to consult additional sources of information such as the data model of the *Discount* object or the possible parameters of the *Apply discount* task to learn more about possible values. However, this is beyond the scope of this paper.

5   Evaluation

In this section, we describe an implementation and an empirical evaluation of our concept. We implemented an efficient and scalable algorithm to check the soundness of decision-aware business processes in *dmn-js*, a DMN decision table editor developed by Camunda. Our implementation uses some of the functionality implemented in the DMN decision table verification algorithm described in [4], namely to determine the output set of a decision table as described in Sect. 4.1. Calvanese et al. report that their algorithm is scalable. Since checking the soundness criteria described in Sects. 4.2 and 4.3 does not introduce additional complexity—the complexity is given by the number of possible outputs of the table times the number of branches of the decision fragment—our algorithm is guaranteed to have the same scalability that is reported in [4].

Since for checking the soundness of decision-aware business processes, it is useful to have a modeler at hand that supports both, process modeling and decision modeling in the same application, we integrated our code into the *Camunda Modeler*[3], a standalone application for the integrated modeling of BPMN and DMN diagrams. You can download our extended, ready to execute *Camunda Modeler* at https://bpt.hpi.uni-potsdam.de/Public/BpmnDmnSoundness*.

Our empirical evaluation investigates the practical relevance of our notion of soundness of decision-aware business processes. The objective was to analyze if modelers indeed make the kinds of errors that Definitions 8 and 9 would catch. For the evaluation, we analyzed two model collections containing associated process and decision models. The first collection we gathered from participants of a massive open online course (MOOC) on process and decision modeling with BPMN and DMN offered by *openHPI* [15] in Spring 2016.[4] The second comes from an implemented process management project of a large German insurance company that deals with automatically handling invoices submitted by customers with private health insurance.

The MOOC lasted six weeks and had around 5500 participants in total. The majority was aged between 30 and 50 years working as consultants with an IT background, while the second most frequent group came from academia. The course required the successful completion of weekly theoretical and modeling exercises as well as a final exam. Additionally, in the last week of the course, we asked the participants to voluntarily submit their solution to a modeling exercise in which they were asked to design a BPMN process model associated with a DMN decision model based on the textual description of a simple billing process, involving a decision about what type of bill to send to the customer. In total, we received 157 unique submissions, 70 of which were appropriate for checking decision-aware soundness. The two most common reasons for the other 87 submissions to be unsuitable for our evaluation were that either the participants only submitted a DMN model but no associated BPMN model (52 submissions), or that the associated BPMN model did not contain a decision fragment as defined

---

[3] https://github.com/camunda/camunda-modeler.
[4] https://open.hpi.de/courses/bpm2016.

in Definition 5 (24 submissions). Rather, the process was designed like in Fig. 10. The original exercise as it was given to the course participants can be accessed at https://bpt.hpi.uni-potsdam.de/Public/BpmnDmnSoundness/Evaluation.

Table 1 shows the results of our analysis. The "cleanest" solution to this exercise required a multi-hit table, but using single-hit tables was also correct and the numbers suggest that the participants favored them over multi-hit tables. Looking at the ratio of unsound processes, in both cases mistakes are made: 71% of the single-hit and 48% of the multi-hit tables were found to be unsound. Thus, in total, 61% of the analyzed tables violate at least one of the soundness criteria of Sect. 4. The most common reason was that the participants used non-matching identifiers for the outputs of the table and the edge conditions of the process.

Table 1. Number and percentage of sound and unsound decision-aware processes

	Sound	Not Sound	Total
Single-hit	12 (29%)	29 (71%)	41 (100%)
Multi-hit	15 (52%)	14 (48%)	29 (100%)
Total	27 (39%)	43 (61%)	70 (100%)

As mentioned above, we also analyzed a BPM project of an insurance company for violations. The project is realized with the *Pega* BPM suite, which is considered a leading vendor of BPM software.[5] Pega defines its own process modeling language that is very closely related to BPMN. Furthermore, it allows XOR-gateways to be associated with decision models such as decision tables. The insurance company applies Pega to realize a BPM project for the automatic processing of claims by privately insured patients. This project contains 86 decision-aware business processes, where each of the associated decision models is a single decision table. Out of these 86 processes, 26 (30%) violate at least one of the decision soundness criteria and are therefore not sound.

We consider the analysis results from both model collections, the MOOC exercise and the Pega project, as important findings of our evaluation, because they show that soundness violations of decision-aware business processes occur frequently. Therefore, tool support for warning the modeler during design time of violations as realized by our implementation is highly desirable. This is especially true in practical settings, such as the one of the insurance company, in which such errors can lead to loss in revenue and reputation.

6 Related Work

Classical soundness checking has been described in [2]. Since this notion is rather strict, various weaker notions have been proposed which are nicely summarized

---

[5] https://www.pega.com/de/bpm.

in [1]. These notions focus on control-flow only and omit interaction, data and resources. Yet, since data plays an important role in process models and execution constraints may exist between activities and data objects, [10] deals with checking conformance between process models and data object life cycles. As in our work, the traditional soundness criteria are deemed insufficient for this endeavour. Therefore, they extend the existing control-flow mappings of process models to Petri nets [8] to also take data flow into account. This enables soundness checking of control flow *and* data.

Employing rules or decision tables for making decisions usually requires considering two aspects of verification: consistency and completeness. Consistency checks are, for example, concerned with ensuring that there are no conflicting rules, i.e., rules that have the same input conditions but different outputs. Completeness checks deal, for instance, with verifying that all possible inputs are assigned an output, which we also require in our criteria. Early works on checking consistency and completeness of rule-based expert systems are described in [6,11,12], where the latter focuses on decision tables. Parts of the algorithms of our paper are based on recent work on the verification of DMN decision tables [4], which focusses on finding missing and overlapping rules.

Recently, a book on the joint use of process and decision modeling in BPMN and DMN was published [7]. This book considers the interaction of processes and decisions. For example, it describes different categories of how BPMN processes may respond to decisions. Nevertheless, there is no discussion of how consistency can be ensured in those categories. In [3,9], it is argued that expressing decision logic in BPMN leads to process models that are hard to read and maintain. Therefore, decision logic should be outsourced. In [9], the authors recommend the use of business rules alongside process models and identify the need for ensuring consistency between the two. Thus, they propose best practices for the integration of rules and processes. Similarly, in [5], a framework is described to integrate BPMN process models and SBVR rules.

7   Discussion and Conclusion

In this paper, we presented an approach for verifying the soundness of decision-aware business processes. These processes separate the concerns of process and decision logic into two different models, process and decision models respectively. With respect to this situation, we defined the notion of decision-aware soundness along with corresponding verification criteria. As already discussed in Sect. 4, our approach is suited for decision models that show what outputs are produced by the decision, and for process models that handle the decision outcome explicitly by conditions annotated at the outgoing branches of the split gateway following the decision. For this reason we also presented a method to determine the set of possible outputs of any kind of DMN decision table.

An interesting direction for future research is to analyze if the set of possible outputs of a decision table depends on the context of the process it is called from. It might be the case that a process is not even able to supply every possible input value combination to the decision table. Hence, requiring the process to cover every output of the decision table that is generally possible might be too strict. Rather, one should check only those outputs that can be produced in the context of the particular process.

We consider our approach as highly relevant because DMN models will be increasingly used in the future in combination with BPMN models to benefit from a separation of concerns [3]. Tool vendors, such as Signavio, are already supporting the modeling of both process and decision models. However, the two models are typically designed independently from each other. This can lead to inconsistencies in different stages. As we showed in our evaluation, soundness violations occur not only during the initial design of the models, but also in implemented models that are already applied in real-world settings. The possibility of applying our soundness checks supported by our extended camunda modeler in these cases will prevent such mistakes.

References

1. van der Aalst, W.M.P., van Hee, K.M., ter Hofstede, A.H.M., Sidorova, N., Verbeek, H.M.W., Voorhoeve, M., Wynn, M.T.: Soundness of workflow nets: classification, decidability, and analysis. Formal Aspects Comput. **23**(3), 333–363 (2010)
2. van der Aalst, W.M.P.: Verification of workflow nets. In: Azéma, P., Balbo, G. (eds.) ICATPN 1997. LNCS, vol. 1248, pp. 407–426. Springer, Heidelberg (1997). doi:10.1007/3-540-63139-9_48
3. Batoulis, K., Meyer, A., Bazhenova, E., Decker, G., Weske, M.: Extracting decision logic from process models. In: Zdravkovic, J., Kirikova, M., Johannesson, P. (eds.) CAiSE 2015. LNCS, vol. 9097, pp. 349–366. Springer, Cham (2015). doi:10.1007/978-3-319-19069-3_22
4. Calvanese, D., Dumas, M., Laurson, Ü., Maggi, F.M., Montali, M., Teinemaa, I.: Semantics and analysis of DMN decision tables. In: La Rosa, M., Loos, P., Pastor, O. (eds.) BPM 2016. LNCS, vol. 9850, pp. 217–233. Springer, Cham (2016). doi:10.1007/978-3-319-45348-4_13
5. Cheng, R., Sadiq, S., Indulska, M.: Framework for business process and rule integration: a case of BPMN and SBVR. In: Abramowicz, W. (ed.) BIS 2011. LNBIP, vol. 87, pp. 13–24. Springer, Heidelberg (2011). doi:10.1007/978-3-642-21863-7_2
6. Cragun, B.J., Steudel, H.J.: A decision-table-based processor for checking completeness and consistency in rule-based expert systems. Int. J. Man Mach. Stud. **26**(5), 633–648 (1987)
7. Debevoise, T., Taylor, J.: The MicroGuide to Process and Decision Modeling in BPMN/DMN. CreateSpace Independent Publishing Platform (2014)
8. Dijkman, R.M., Dumas, M., Ouyang, C.: Semantics and analysis of business process models in BPMN. Inf. Softw. Technol. **50**(12), 1281–1294 (2008)
9. Hohwiller, J., Schlegel, D., Grieser, G., Hoekstra, Y.: Integration of BPM and BRM. In: Dijkman, R., Hofstetter, J., Koehler, J. (eds.) BPMN 2011. LNBIP, vol. 95, pp. 136–141. Springer, Heidelberg (2011). doi:10.1007/978-3-642-25160-3_12
10. Meyer, A., Weske, M.: Weak conformance between process models and synchronized object life cycles. In: Franch, X., Ghose, A.K., Lewis, G.A., Bhiri, S. (eds.) ICSOC 2014. LNCS, vol. 8831, pp. 359–367. Springer, Heidelberg (2014). doi:10.1007/978-3-662-45391-9_25

11. Nguyen, T.A., Perkins, W.A., Laffey, T.J., Pecora, D.: Checking an expert systems knowledge base for consistency and completeness. In: Proceeding of IJCAI, pp. 375–378. Morgan Kaufmann Publishers Inc. (1985)
12. Nguyen, T.A., Perkin, W., Laffey, T.J., Pecora, D.: Knowledge base verification. AI Mag. 8(2), 69–75 (1987)
13. OMG: Business Process Model and Notation, Version 2.0.2, January 2014
14. OMG: Decision Model and Notation, Version 1.1, May 2016
15. Totschnig, M., Willems, C., Meinel, C.: openHPI: evolution of a MOOC platform from LMS to SOA. In: Proceeding of CSEDU, pp. 593–598. SciTePress (2013)
16. Vanthienen, J., Dries, E.: Developments in decision tables: evolution, applications and a proposed standard. DTEW Research Report, KU Leuven (1992)
17. Von Halle, B., Goldberg, L.: The Decision Model: A Business Logic Framework Linking Business and Technology. Taylor and Francis Group, Boca Raton (2010)
18. Weske, M.: Business Process Management: Concepts, Languages, Architectures, 2nd edn. Springer, Heidelberg (2012)
19. Xylander, O.: On the Relationship between Decision Modeling and Process Modeling. Master's thesis. Hasso Plattner Institute, Potsdam, Germany (2015)

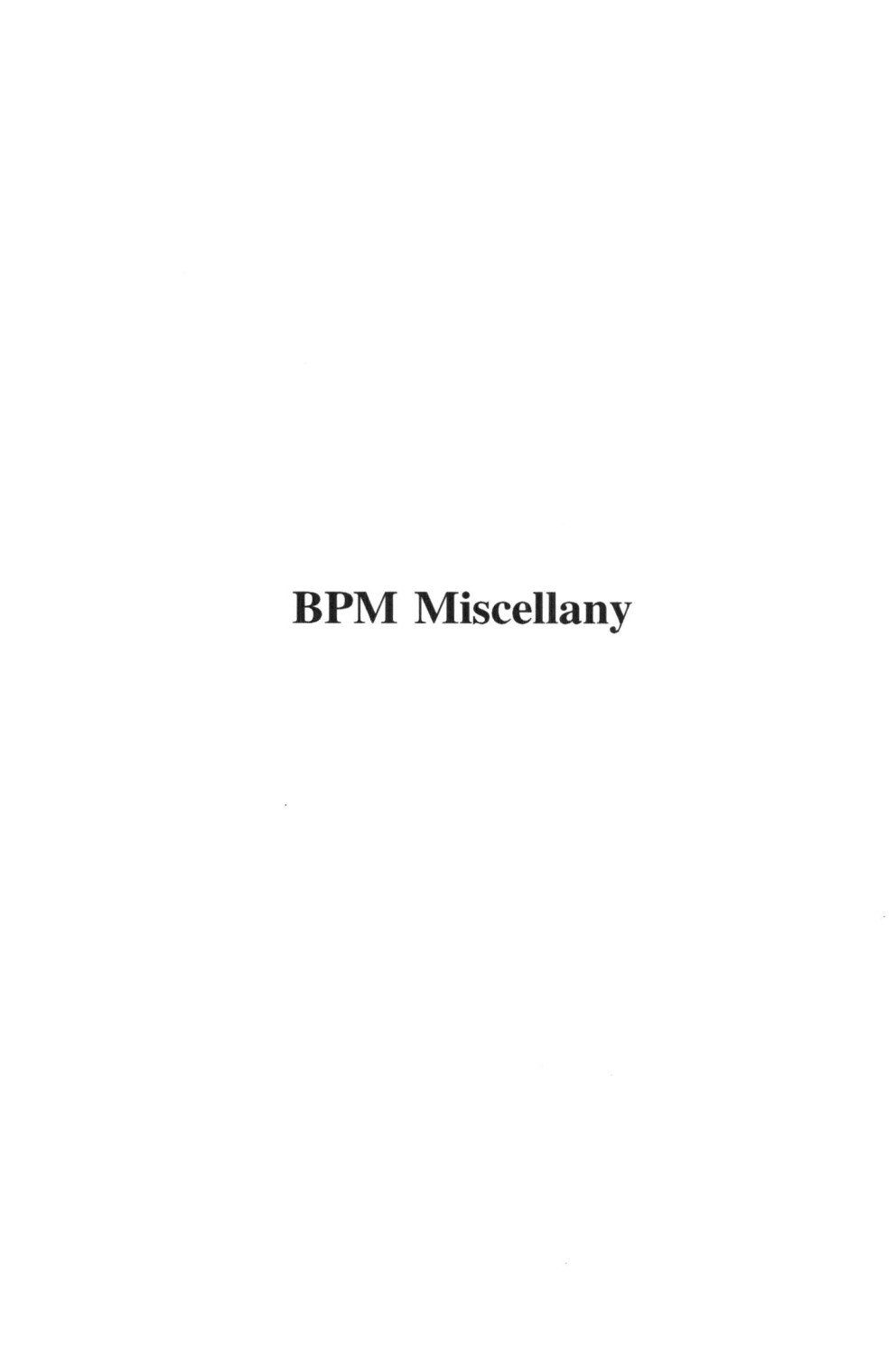

BPM Miscellany

BPMS-Game: Tool for Business Process Gamification

Javier Mancebo[1(✉)], Felix Garcia[1], Oscar Pedreira[2],
and Maria Angeles Moraga[1]

[1] Institute of Technology and Information Systems,
University of Castilla-La Mancha, Ciudad Real, Spain
{javier.mancebo,felix.garcia,
mariaangeles.moraga}@uclm.es
[2] Database Laboratory, University of A Coruña, A Coruña, Spain
opedreira@udc.es

Abstract. Recent years have seen an increase in the concern for the environment, with attempts being made to raise public awareness about the need to have sustainable development that enables the use of any resource in the present, without compromising its future. This concept of sustainability should be applied in the lifestyle of individuals, but also in companies or organizations that deal with business process management, known as BPM. This would allow the creation of business processes that are more sustainable and which use resources more efficiently; this is the concept of Green BPM. It is thus of prime importance to find an incentive for the workers in the companies to get involved in these sustainable initiatives. This has led to a consideration of the need to incorporate gamification, i.e., the use of game elements in non-game contexts in an effort to induce certain behaviors in people. The aim of these games would be to enhance participation and foster commitment to sustainable development. With all these issues in mind, in this article the BPMS-Game tool is described; the tool combines the concepts of gamification, sustainability, and business processes to support the definition of games that promote sustainability in BPM environments. A set of base and derived measures have been defined to evaluate the user behavior with respect to sustainability in their daily work when using a BPMS system. The contributions that BPMS-Game can offer are illustrated with a representative example.

Keywords: Business processes · Gamification · BPMS · Sustainability · Green BPM

1 Introduction

The lifestyle of our present-day society is threatening the future existence of the resources we employ. This makes it necessary to promote sustainable development, thus avoiding endangering the environment. We define sustainability as the ability to employ any resource at the present time without compromising the use of this resource for future generations to be able to meet their own needs, thereby ensuring a balance between economic growth, social welfare and the environment [1]. All individuals and

© Springer International Publishing AG 2017
J. Carmona et al. (Eds.): BPM Forum 2017, LNBIP 297, pp. 127–140, 2017.
DOI: 10.1007/978-3-319-65015-9_8

organizations really need to become aware of the importance of action in this area, and they should get involved in the quest to achieve a high degree of sustainability. If we focus on the information technology industry, different initiatives have appeared in recent years; their goal has been to improve energy efficiency, and consequently achieve greater sustainability [2]. These trends have led to a current area of action that is known as Green IT; this is based on cost reduction, and focuses on issues that adversely affect the environment, both in the design and construction of computer systems [3].

These Green IT trends can have a great influence on the paradigm of BPM, which strongly relies on the use of Information Technologies, and is focused on efforts to optimize the organization's business processes. In an effort to improve productivity, effectiveness and efficiency of systematic management processes, these have to be modeled, automatized, integrated, monitored and optimized continuously [4]. Business process management can thus contribute to Green IT initiatives, facilitating these and helping to create business processes that are more sustainable and more efficient as regards the resources employed [2, 5]. The synergy between BPM and Green IT has given rise to the term Green BPM, also known as sustainable management of business processes [6]. The human factor is one of the main elements of success as regards a proper implementation of the Green BPM tendency in the organization [7, 8]. This means that it is essential to consider the influence that people and their interactions can have on business processes, and take into account the importance of their being involved in sustainable initiatives.

This article tackles the above issues, looking at the potential benefits that can be obtained by improving sustainability in a business environment and addressing these from the perspective of the human factor. It does this with special reference to companies or organizations that have evolved in recent years towards developing business process management. In that sense, it has been seen that the potential impact of a proper application of the concept of gamification in promoting sustainability should be highlighted, since this approach has indeed experienced tremendous growth in recent years. Gamification can be defined as the "use of elements of game design in the context of non-game" [9]. To put it another way, it is the application of thoughts and game mechanics in more serious environments to induce certain behavior in people who are interacting with the game. At the same time, it seeks to improve the participation, motivation and commitment of a user while performing a particular task. We might also say that gamification takes those characteristics that make games fun and attractive (and even addictive) and uses them to improve the player experience in a non-gambling environment, such as those of business or education [10].

The purpose here, then, is to employ gamification with game mechanics and dynamics and to attempt to use these to motivate workers of an organization to follow a series of green initiatives in the business processes they interact with. The main objective of the work presented in this article is the development of an environment that promotes business processes with a higher level of sustainability by encouraging users of BPMS platforms to be more environmentally friendly in their daily work. The tool analyzes the logs of BPMS systems and tries to engage users, motivating them to be more sustainable in their work by means of gamification mechanisms.

The rest of the paper is organized as follows: In Sect. 2, the related work is set out, including the background of the topics addressed. Section 3 presents the proposal for the measures that are used to evaluate the behavior of users with regard to sustainability. The BPMS-Game tool is explained in Sect. 4, where an example of the use of the tool developed is presented. Lastly, the main conclusions and proposals for future work are put forward in Sect. 5.

2 Related Work

We have found that there is a certain lack of studies addressing the notion of combining the concept of sustainability and business process management so that Green BPM [11] comes about as a result. To date, the vast majority of relevant studies deal with sustainable IT projects. Recker et al.'s work [6] shows the importance of measuring the sustainability in business processes, and introduces an analysis method for measuring the carbon dioxide emissions produced by the execution of a business process. If we focus on the sustainability of information technology, many of these studies aim to measure the sustainability of the processes solely in the hardware infrastructure. Aleksic [12] points to the need to measure the energy consumption of the IT infrastructure used, showing the different activities that can be used to save energy, thereby improving Green IT. Betz and Caporale [13], for their part, focus on the importance of measuring the sustainability of the software applications that are used. Another important work is that of Cappiello et al. [14], which shows an approach that promotes an efficient use of energy by the design of processes that have low energy consumption.

Concerning the measures, the authors divide these into several different types: according to the amount of energy consumed, (which is the most popular measure), the number of emissions generated, the number of raw materials or resources used, and other types of measures [2, 5, 6, 11, 12, 15]. In this context, Welter et al. [16] have carried out a systematic mapping on the green metrics that can be used by organizations that are responsible for software development.

The term sustainability has grown in popularity, as has another term, i.e. gamification; the latter has attracted significant attention in various domains such as mobile applications or education, among others [17]. Several authors have shown the advantages and benefits of employing gamification, such as Hamari et al. [18], through a literature review. It was concluded in this work that the application of gamification actually works, but it is really necessary to take certain caveats into account. For instance, gamification is based on very basic game mechanics in some cases (points, levels or classifications); more advanced aspects, such as social interaction or mobility issues, should also be taken into consideration.

In recent years, a large proportion of the research work has focused on how to apply gamification in software engineering environments. It is a considerably young line of research because, as can be seen in a systematic mapping [10] of existing studies in the field of gamification in software engineering, the first articles date only from 2010. Regarding the type of game elements and the mechanics applied to the existing proposals, award-badge systems based on points were the most relevant, followed by the rankings of classification, social elements, and dashboards. There are at present a number of different commercial tools that provide support to the software engineering process by incorporating the basic mechanisms of gamification mentioned above; some examples are RedCritter, PropsToYou, ScrumKnowsy, Masterbranch, Co-deHunt, The Continuous Integration Game, or the plugin Jenkins, among others. There are also some gamification platforms which, applied together with corporate tools of an organization, help to create a gamification environment. These include platforms of general gamification such as: Badgeville (www.badgeville.com); Gamify - (www.gamify.es); Bunchball Nitro - (www.bunchball.com/nitro).

3   Proposal of Sustainability Measures for Business Processes

The key aspect of the gamification environment provided by BPMS-Game is a suitable evaluation of the behavior of users, in an effort to promote more sustainability in their work. To achieve this, first of all various entities which may be involved for each of the types of tasks described in the BPMN standard [19] have been identified, since, depending on the type of task that is being executed, one type or another can be applied. For example, for the type of tasks related to sending and receiving, the different IT resources (computers, communication devices, etc.) have been identified as the main entities. For manual tasks, a number of different entities have been identified, apart from IT resources. These include software applications other than the BPMS environment, manufacturing machines that are used in the process, or vehicles that may be needed for transportation of either people or goods in the execution of a task in a process. Once the entities that might be involved for each of the types of tasks described in the BPMN standard are identified, it is necessary to define what measures can be used to determine the sustainability of the participant entities. We divide the proposal of the measures selected into two types: on the one hand, the base measures, which are measures of an attribute that do not depend upon any other measure, and whose measurement approach is a measurement method; and on the other hand, the derived measures, which are the measures derived from other base or derived measures, using measurement functions as measurement approaches [20].

The base measures have been extracted from a systematic literature review whose main aim was to identify the entities, attributes, and measures that can be used to evaluate the sustainability of the business processes. This revision includes primary studies published between 2010 and 2016. Table 1 shows a list of the base measures that have been selected from the primary studies with reference to the attributes considered.

In addition, a proposal for derived measures—which allows us to assess the sustainability of the execution of a specific case or a complete implementation process—has been formulated. The derived measures are those coming from other base measures using a measurement function. In other words, a derived function is an algorithm or a calculation based on combining two or more base or derived measures [20]. Table 2 shows the resulting list of measures proposed to assess sustainability. In the example of the use of the application, (Sect. 4) how these measures are employed will be explained in detail.

Table 1. List of measures extracted from a review of the existing literature on sustainability measures.

Attribute	Measure	Comments	Source
Energy	Power consumed (W)	Amount of power needed to operate the corresponding entity	[2, 12, 13, 15, 21]
	Energy consumed (kWh)	Amount of energy consumed per hour of work	[13–15, 21]
	PUE (Power usage effectiveness)	The result of dividing the total power of a CPD between the power available for the computer equipment	[22]
	DCE (Data Center Infrastructure Efficiency)	The inverse of the PUE	[22]
	DCeP (Data Center energy Productivity)	It serves to quantify the useful work produced compared to the energy required	[22]
	Maximum kWh	Maximum amount of kilowatts (kW) consumed in one hour	[14]
	Work done/energy consumed	A derived measure used to measure energy efficiency	[21, 23]
Emissions	g. CO_2/h	Measures the amount (in grams) of carbon dioxide (CO2) per hour of run	[5, 13, 22, 24]
	g. CO_2/Kg paper	Amount of CO2 (in grams) produced per kg. of paper used	[25]
	g. CO_2 by km	Amount of CO2 (in grams) produced per km. traveled using a vehicle with fuel	[5, 25]
	g. CO_2/kWh	Amount of CO2 (in grams) produced by generating each kilowatt hour spent	[25]
Consumption of resources or raw materials	Kg of paper	Amount (in kilograms) of paper used in each task	[5, 13, 15, 24]
	l. of water	Amount (in liters) of water used in each task	[5, 24]
	l. of fuel	Amount (in liters) of fuel used in each task	[5]
	l. Printing ink	Amount (in liters) of printing ink used in each task	[5, 15]
Waste	Amount of toxic materials	Amount of toxic materials discarded when a task is carried out	[13, 15]
	Number of discarded electrical devices	Amount of electrical devices (computers, printers …) discarded due to their obsolescence or breakage over a given period of time	[13, 15]
Software	Number of lines of code (LOC)	An estimate of lines of code, which can affect to the amount of energy consumed by an application	[23]
	Number of Loop Cycles	It can affect to the amount of energy consumed by an application	[23]

Table 2. List of derived measures proposed to assess sustainability

Id	Derived measures	Comments
Md1	Total power required in a BP case (Md1 = Σ Watt (W) per task)	The sum of all the powers necessary for the implementation of each of the tasks of a business process case
Md2	Total energy consumed in a BP case (Md2 = Σ kWh per task)	The sum of all the energy consumed for the execution of all tasks belonging to a BP case
Md3	Activity that consumes most energy (kWh) in a BP case	The task executed that consumes most energy in a BP case
Md4	Activity that consumes most energy (kWh) on average in a BP case	The task of a BP case that consumes most energy on average of all executions in that case
Md5	Work done/energy consumed of each of the tasks	Energy efficiency resources consumed by each of the tasks of a BP
Md6	Overall energy efficiency of a BP case (Md6 = Σ Md5)	The overall efficiency of a BP case, taking into account the efficiency of each of the tasks executed
Md7	Activity with the highest energy efficiency in a case	The task executed in a BP case has the greatest energy efficiency
Md8	Quantity emissions (CO 2) generated in a BP case (Md8 = Σ $g.CO_2$ per task)	The sum of emissions of CO 2 in the execution of a BP case
Md9	Activity that generates the greatest amount of CO 2 in a BP case	The task that generates the greatest amount of emissions among those executed to finalize a BP case
Md10	Activity that generates the greatest amount of CO2 on average in a BP case	The task of a BP case generating most emissions on average of all executions in the case
Md11	Amount of fuel used in a BP case (Md11 = Σ l. fuel per task)	The sum of the amount of fuel consumed in the execution of all tasks belonging to a BP case
Md12	Amount of ink used in a BP case (Md12 = Σ ml of ink. per task)	The sum of the amount of printing ink spent in the execution of all tasks belonging to a BP case
Md13	Amount of water used in a BP case (Md13 = Σ l. water per task)	The sum of the amount of water used in the execution of all tasks belonging to a BP case
Md14	Amount of paper used in a BP case (Md14 = Σ kg. Paper per task)	The sum of the amount of paper used in each of the tasks of an executed BP case
Md15	Average energy consumed in BP cases	The calculation of the average power consumed in the execution of all BP cases
Md16	Average emissions generated in BP cases	The calculation of the average emissions generated in the execution of all BP cases
Md17	Average energy efficiency of the cases in a BP	The average energy efficiency of the execution of all BP cases
Md18	Average raw material used in a BP	The amount of average raw materials used in the execution of all BP cases
Md19	Activity that has the highest energy efficiency in a BP	The task executed in a BP that has the greatest energy efficiency
Md20	Most energy-consuming activity of a BP	The task executed in a BP that consumes most energy
Md21	Activity that generates most emissions in a BP	The task executed in a BP that generates most CO2 emissions

4 BPMS-Game Tool

BPM provides organizations with a tool for automating and managing their processes, usually by monitoring and optimizing variables related to time, effort, cost, and quality. In addition to those traditional management variables, we can also use process variables related to sustainability of the processes. In this section, the BPMS-Game is described, firstly by presenting its main features in the application of gamification to business process management. We developed this tool with the goal of improving process sustainability based on the relevant information that can be extracted from the execution logs in BPMS systems. The potential contribution of the tool is also illustrated by means of an example of its application.

The main functionalities that the BPMS-Game tool supports are:

- The BPMS-Game tool is applicable to any BPMS system, thus enabling the information on the execution of business processes from execution logs, defined in the standard format XES (Extensible Event Stream), to be obtained. General information about the process and each of the tasks executed is extracted from each of the logs processed by the tool. This includes information on the elements that allow gamification to take place, namely the name of the task, the human resource executing the tasks (if not run automatically), the number of tasks, or the time it takes to complete them, taking performance indicators into consideration. Information concerning the sustainability of each of the tasks that have been executed is also extracted. This information constitutes the base measure on which to define the indicators that are to be evaluated in the rules of the game.
- The application allows the management of users participating in the system; these are the human resources taking part in the project.
- BPMS-Game automatically performs a calculation of the base measure with the information about the sustainability of tasks that is used afterwards to create rules.
- The tool should allow the creation of the rules of the game, as well as achievements established for each of the rules.
- Once the objectives have been achieved, the application will allow users to redeem their achievements for rewards or gifts, and will provide a visual display of the progress achieved in the game.

Figure 1 shows the UML diagram of the domain model of the application. As can be seen in the domain model, the BPMS-Game application is composed mainly of the processes and their cases, which can be extracted from the execution logs of a BPMS, along with the users that participate in it, either as managers or players, and the rules that are created by the administrator; through a rules engine, the rules are evaluated in real time by continuous monitoring of the execution logs. A player reaches a reward any time he/she fulfills some of the rules defined; these may be in the form of badges, levels or points, which are the mechanics employed to create a gamification environment in the BPMS-Game tool. The points obtained by each of the players can be redeemed for various prizes that the administrator has entered into the system.

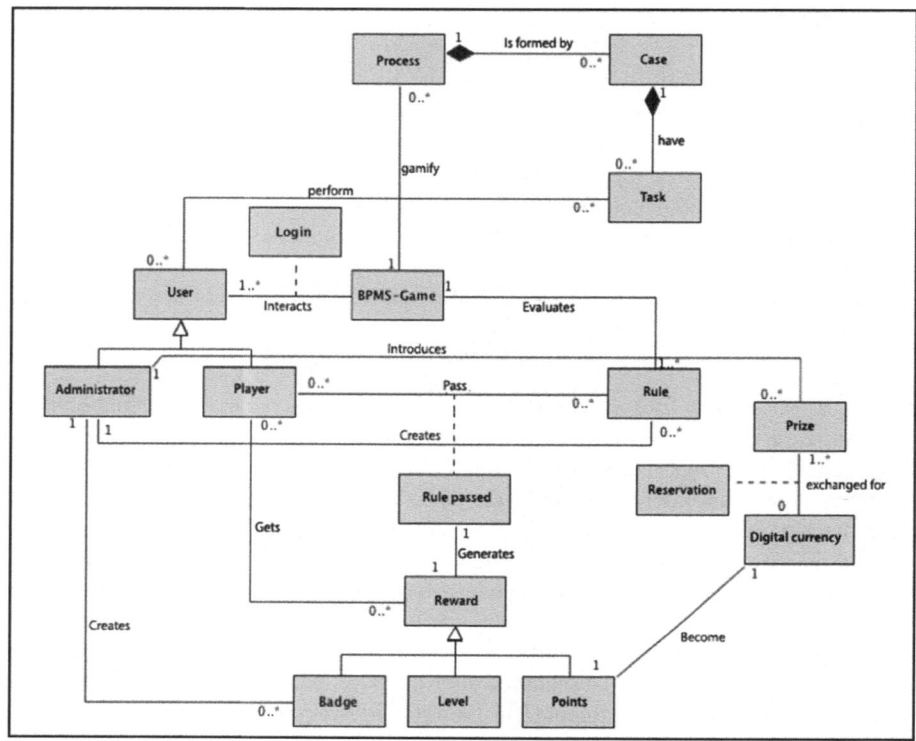

Fig. 1. BPMS-game domain model

The tool should allow the administrator to create badges as mechanics of the implemented gamification. These badges are composed of an image, a code (which must be unique for each of the badges) and a representative name.

BPMS-Game allows the management of the different prizes that players can obtain according to the points they are getting. To that end, the Administrator must create awards; these awards may also be modified or deleted. An attribute of awards is the Value; this value is defined in GCoins, which is a virtual currency created so that users can redeem their points for the prizes that the administrator has previously entered into the system. In addition to all the features already mentioned, an option has been added whereby users can share general game information, such as their points and the level they have reached on trending social networks (Facebook, Twitter, and Google+).

4.1   BPMS-Game: Example of Application

In this section, an example of application, which serves as a proof of concept of the potential utility of BPMS-Game, is described. The scenario chosen to illustrate the sustainability assessment of a business process is a process for picking up patients from

Table 3. Simulated data for Task 2 and 4.

Task	Entity	Measure	Data
Task 2	Tablet	Consumed energy (kWh)	0.050 kWh
		g. CO_2/kWh	33 g CO_2
	Software application	Energy impact on resource (kWh)	0.015 kWh
Task 4	Ambulance	g. CO_2 km	800 g CO_2 km
		l. of fuel	0.330 l Km

care homes for the elderly to take them to the hospital. Figure 2 shows the BPMN2 model of the business process.

In this process, three different roles can be identified: the nursing assistant, who is responsible for giving the order to pick up patients; the ambulance driver, who must evaluate the requests and pick up the patients to then drive them to the hospital; and the orderly, who fills in the pick-up document of the patients. The patient pick-up process includes seven tasks: three belong to one kind of user, two are manual, one is a call to a service, and there is another task.

It is necessary to know which particular entities are involved in each of the tasks defined in the process. For example, for Task 1, the entities identified are the hardware resources used (presumably the computer and a printer) and the software responsible for generating the patients' documentation. For tasks 4 and 5, which are of a manual type, the only entity identified is the ambulance used to transport patients.

Once the entities in all the tasks of the business process are identified, the measurement of the sustainability of each of these must be carried out using the measures

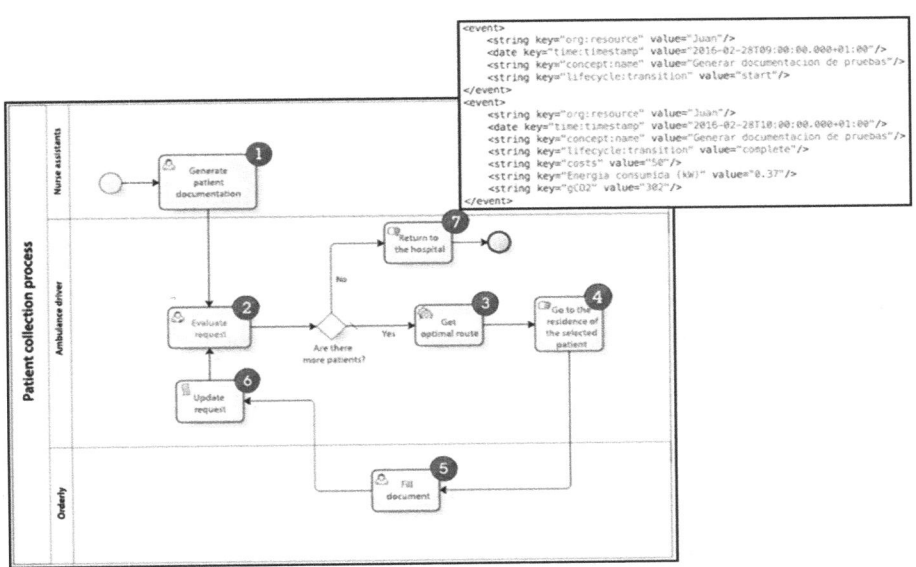

Fig. 2. Application example of BPMS-Game.

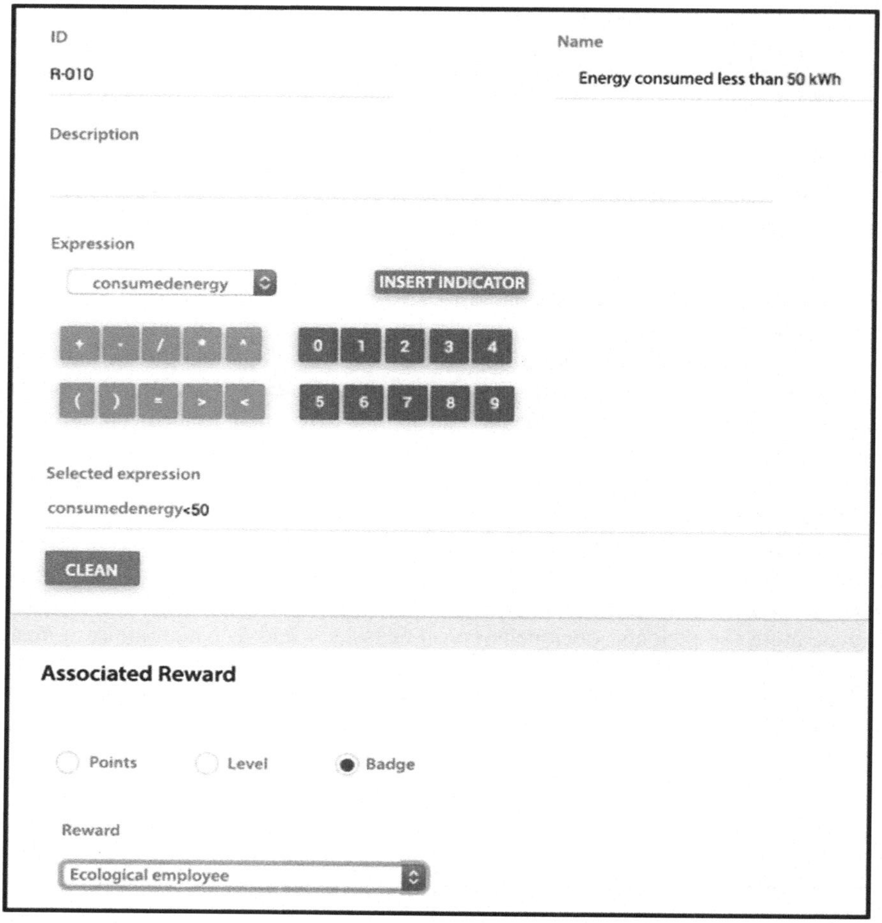

Fig. 3. BPMS-Game: Creation of a new rule.

defined above. The measurement process has been simulated for this example using fictitious data which are representative to calculate the proposed measures. Table 3 shows the data of the measurements for tasks 2 and 4.

Once all the data have been simulated for each of the process tasks, the sustainability of a specific execution case can be evaluated, using for that purpose the derived measures that are shown in Table 3.

We evaluate the sustainability of a concrete case (Fig. 2) below, where the following tasks are performed: Task 1 – Task 2 – Task 3 – Task 4 – Task 5 – Task 6 – Task 2 – Task 7.   For this example, we will use only measures about the energy consumed, as well as the quantity of CO_2 generated. When we have data on the measures of each of the tasks, we can calculate the results of the chosen measures.

- Total consumed energy (Md1) = Σ Consumed energy (Task i):
  Md1 = 0.37 + 0.065 + 0.365 + 0.065 + 0.37 + 0.365 + 0.065 + 0.065 = 1.73 kWh
- Total CO2 generated (Md8) = Σ CO2 generated (Task i):
  Md9 = 33 + 10000 + 33 + 302 + 12000 + 33 + 33 + 302 = 22736 g. CO2

In Fig. 2 a short excerpt of the execution log is displayed. One of the problems encountered when evaluating BPMS-Game is that most BPMS tools currently represent the logs in a proprietary format, so it was necessary to generate the log in XES from the process execution manually in several cases with multiple users, using the functionality offered by some BPMS to monitor the execution of the processes. XES log files already contain the general information about the process, as well as the sustainability information of each of the tasks that the resources have executed. Once the system has processed the information of XES logs, the administrator is responsible for creating users on the BPMS-Game platform with their profile information, along with the badges that users can obtain as rewards and the prizes that they can redeem.

The administrator can also create the rules of the game that will subsequently be evaluated by the tool automatically, assigning rewards to the users, rewards that are associated with success in the challenges presented by the rules. To define the rules, the administrator may use performance indicators (number of completed tasks) or sustainability indicators, such as energy consumed in completed tasks. Figure 3 shows the creation of a new rule, using sustainability indicators, for employees that consumed less than 50 kWh in the execution of their tasks.

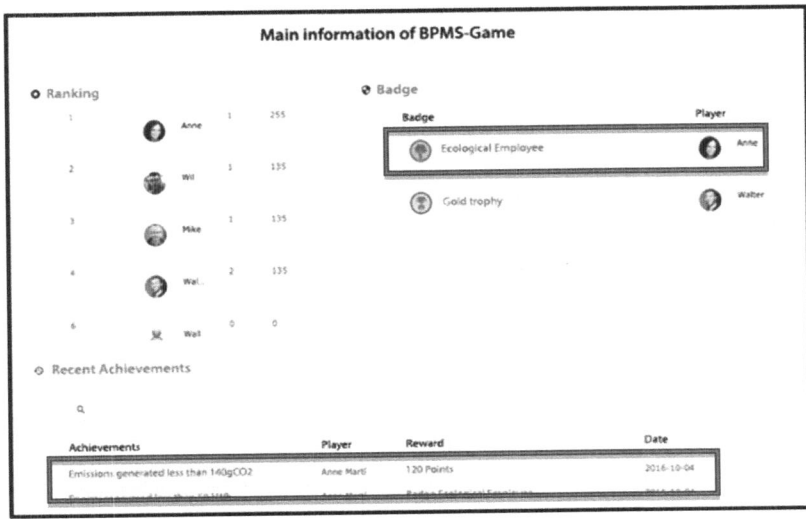

Fig. 4. BPMS-game: ranking and achievements.

Once the administrator has created the sustainability rules, the application is automatically responsible for evaluating these rules; it allocates rewards to each player who has been successful in fulfilling the rule, thereby performing gamification. Figure 4 displays some players who have been successful in performing according to the two rules related to sustainability that had been created previously by the administrator.

As can be seen in the example, the BPMS-Game tool includes the different derived measures that have been defined in the proposal, and which allow a gamification of the users to be carried out using the indicators of sustainability.

5   Conclusions and Future Work

The aim of this paper is to establish a proposal of different measures for measuring and improving the sustainability of business processes. To achieve that goal, an analysis of the existing sustainability measures in the related literature has been carried out. Using the measures found, a set of derivate measures that can evaluate the sustainability is proposed. It is important to highlight that these measures consider the different entities that are involved in the business process.

The BPMS-Game has been presented in this work, as a tool that supports the definition of games on BPMS platforms and enables their evaluation from execution logs of such platforms in the quest to improve the sustainability of business processes.

The contribution that BPMS-Game offers in the field of BPM is that by introducing gamification alongside the concept of sustainability, user involvement is promoted and this motivates them to participate in business process tasks. Furthermore, it encourages users to be more environmentally friendly in their daily work, remembering that people are one of the main components on which BPM and its BPMS applications are based.

The intention—for future work— is to define a set of indicators that serve as an analysis model to assess whether the sustainability result obtained is acceptable or not, or to give a degree of its compliance. Likewise, with respect to sustainability in business processes, the intention is to create an energy classification model of business processes, similar to those used in other fields, such as housing or appliances, allowing the assignment of a quantitative mark to business processes according to the evaluation of their sustainability.

On the other hand, our next goal is to validate and evaluate, both the proposed measures and the BPMS-Game tool, in a real setting and thus to know how gamification can improve specific BPM practices.

Acknowledgements. This work has been partially supported by the following projects: GIN-SENG (Ministry of Economy and Competitiveness and the European Regional Development Fund ERDF, TIN2015-70259-C2-1-R).

References

1. Brundtland, G.H.: Report of the World Commission on environment and development: our common future, United Nations 1987
2. Houy, C., Reiter, M., Fettke, P., Loos, P.: Towards green BPM – sustainability and resource efficiency through business process management. In: Muehlen, M., Su, J. (eds.) BPM 2010. LNBIP, vol. 66, pp. 501–510. Springer, Heidelberg (2011). doi:10.1007/978-3-642-20511-8_46
3. Binder, W., Suri, N.: Green computing: energy consumption optimized service hosting. In: Nielsen, M., Kučera, A., Miltersen, P.B., Palamidessi, C., Tůma, P., Valencia, F. (eds.) SOFSEM 2009. LNCS, vol. 5404, pp. 117–128. Springer, Heidelberg (2009). doi:10.1007/978-3-540-95891-8_14
4. Weske, M.: Business Process Management - Concepts, Languafes, Architectures. Springer, Heidelberg (2007)
5. Hoesch-Klohe, K., Ghose, A., Lam-Son, L.: Towards green business process management. In: 2010 IEEE International Conference on Services Computing (SCC), pp. 386–393 (2010)
6. Recker, J., Seidel, S., vom Brocke, J.: Green business process management. In: vom Brocke, J., Seidel, S., Recker, J. (eds.) Green Business Process Management: Towards the Sustainable Enterprise, pp. 3–13. Springer, Heidelberg (2012)
7. Rosemann, M., vom Brocke, J.: Handbook on Business Process Management 1: Introduction, Methods, and Information Systems, 1st edn. Springer, Heidelberg (2015)
8. Trkman, P.: The critical success factors of business process management. Int. J. Inf. Manage. **30**, 125–134 (2010)
9. Deterding, S., Dixon, D., Khaled, R., Nacke, L.: From game design elements to gamefulness: defining gamification. In: Proceedings of the 15th International Academic MindTrek Conference: Envisioning Future Media Environments, pp. 9–15 (2011)
10. Pereira, O., Garcia, F., Brisaboa, N.R., Piattini, M.: Gamification in software engineering – A systematic mapping. Information and Software Technology (2014)
11. Recker, J.: Green, Greener, BPM? BPTrends **5**, 1–8 (2011)
12. Aleksic, S.: Green ICT for sustainability: a holistic approach. In: 2014 37th International Convention on Information and Communication Technology, Electronics and Microelectronics (MIPRO), pp. 426–431 (2014)
13. Betz, S., Caporale, T.: Sustainable software system engineering. In: 2014 IEEE Fourth International Conference on Big Data and Cloud Computing (BdCloud), pp. 612–619 (2014)
14. Cappiello, C., Fugini, M., Ferreira, A.M., Plebani, P., Vitali, M.: Business process co-design for energy-aware adaptation. In: 2011 IEEE International Conference on Intelligent Computer Communication and Processing (ICCP), pp. 463–470 (2011)
15. Löser, F.: Green IT and green IS: definition of constructs and overview of current practices. In: Proceedings of the 19th Americas Conference on Information Systems (2013)
16. Welter, M., Benitti, F.B.V., Thiry, M.: Green metrics to software development organizations: a systematic mapping. In: 2014 XL Latin American Computing Conference (CLEI), pp. 1–7 (2014)
17. Zichermann, G., Cunningham, C.: Gamification by Desing: Implementing Game Mechanics in Web and Mobile Apps. O'Reilly, Sebastapol (2011)
18. Hamari, J., Koivisto, J., Sarsa, H.: Does gamification work? – a literature review of empirical studies on gamification. In: 2014 47th Hawaii International Conference on System Sciences, pp. 3025–3034 (2014)
19. OMG-BPMN, Business Process Model and Notation (BPMN), ed (2011)

20. García, F., Bertoa, M.F., Calero, C., Vallecillo, A., Ruíz, F., Piattini, M., et al.: Towards a consistent terminology for software measurement. Inform. Softw. Technol. **48**, 631–644 (2006)
21. Moraga, M.Á., Bertoa, M.F.: Green software measurement. In: Calero, C., Piattini, M. (eds.) Green in Software Engineering, pp. 261–282. Springer, Cham (2015)
22. Opitz, N., Erek, K., Langkau, T., Kolbe, L., Zarnekow, R.: Kick-starting green business process management - suitable modeling languages and key processes for green performance measurement. In: Association for Information Systems (2012)
23. Johann, T., Dick, M., Naumann, S., Kern, E.: How to measure energy-efficiency of software: Metrics and measurement results. In: 2012 First International Workshop on Green and Sustainable Software (GREENS), pp. 51–54 (2012)
24. Seidel, S., Recker, J.: Implementing green business processes: the importance of functional affordances of information systems. In: ACIS 2012: Proceedings of the 23rd Australasian Conference on Information Systems (2012)
25. Recker, J., Rosemann, M., Hjalmarsson, A., Lind, M.: Modeling and analyzing the carbon footprint of business processes. In: Green Business Process Management, pp. 93–109. Springer, Heidelberg (2012)

Events in Business Process Implementation: Early Subscription and Event Buffering

Sankalita Mandal[1]([✉]), Matthias Weidlich[2], and Mathias Weske[1]

[1] Hasso Plattner Institute at the University of Potsdam, Potsdam, Germany
{sankalita.mandal,mathias.weske}@hpi.de
[2] Department of Computer Science,
Humboldt-Universität zu Berlin, Berlin, Germany
matthias.weidlich@hu-berlin.de

Abstract. Event handling is a fundamental concept for the implementation of business processes. It enables the specification of how a process communicates with its environment and how this environment influences the execution of a process. However, even feature-rich languages for process specification such as BPMN are severely limited in their event handling semantics. They largely neglect the design choices to be made when deciding on when to subscribe to event sources and how to retrieve events for a particular process instance. In this paper, we therefore propose a model for event handling in business processes that is grounded in explicit subscriptions and event buffering. This model is integrated in BPMN using its extension mechanism and comes with formal execution semantics. Based on the latter, we further show how existing techniques for verification and adapter synthesis can be leveraged to analyse the interactions of a business process. Finally, we demonstrate the feasibility of our event handling model by means of an implementation in Camunda, an open-source process engine.

Keywords: Process implementation · Event processing · Event subscription · BPMN

1 Introduction

The implementation of business processes in state-of-the-art process engines is model-driven: a process model defines a set of activities along with causal and temporal dependencies for their execution [34]. Current languages for process specification such as Business Process Model and Notation (BPMN) [29] feature a wide range of concepts to capture business processes, from basic control-flow dependencies, through advanced handling of events and exceptions, to organisational and data modelling [14].

Nowadays, most business processes are executed in a distributed setting and rely on interactions with their environment. For example, process execution may wait for the completion of tasks by external participants or be aborted based on sensed data (e.g., location information of process-related entities). To

© Springer International Publishing AG 2017
J. Carmona et al. (Eds.): BPM Forum 2017, LNBIP 297, pp. 141–159, 2017.
DOI: 10.1007/978-3-319-65015-9_9

specify these interactions, process specification languages feature event handling constructs. Taking BPMN [29] as an example, production and consumption of events is captured by explicit event constructs, artefacts that define the payload of events, message flows linking to external process participants, and event-based control-flow routing (e.g., event-based gateways, boundary events, or event sub-processes). As such, common process specification languages comprehensively capture *what an event is* and *how it influences the control-flow*.

Taking into account that events are typically generated by sources that are external to a process engine, event handling raises two additional questions, though:

When to subscribe to an event source? External event sources are commonly unaware of the execution of a process. Hence, it needs to be clarified from which point in time an event subscription is issued, i.e., when events of a particular type may become relevant for a process instance. This question is independent of any correlation conditions that determine the relevance of an event for a process instance based on its payload data [6,18].

How to retrieve events for a process instance? Once a subscription has been issued, relevant events are stored by a process engine until they are consumed by a process instance. If more than one event is available, however, the following questions need to be clarified: How many and which events shall be considered when selecting the event to consume? Can an event be consumed multiple times?

Most process specification languages ignore the above questions and, if at all, define simplistic semantics, which cannot capture many important scenarios. We illustrate the resulting issues by means of BPMN, but note that other languages, e.g., UML Activity Diagrams [28] or WS-BPEL [4], share the same limitations. According to the BPMN specification [29], semantics of explicit event constructs are such that: (i) a process instance is ready to consume an event immediately upon activation of the event construct by the control-flow; (ii) the process instance then waits until an event is observed, before continuing execution. Adopting this semantics, one cannot express that a process instance shall consume an event that was emitted by the environment *before* the control-flow reached the event construct. Put differently, the time of event subscription is bound to the time of control-flow activation, thereby limiting the types of interactions between a process and its environment that can be specified in a model.

In this paper, we address the above questions and outline how the concepts of early subscription and event buffering help to overcome the aforementioned issues. The contributions of this paper are summarised as follows:

- *Event Handling Model:* We propose a model to explicitly issue event subscriptions in process models and define retrieval policies for event consumption. This model is proposed as an extension to BPMN.
- *Formal Semantics and Analysis:* We define formal semantics for our model using Petri-nets. We also detail how this formalisation enables analysis of

the interactions of a process based on existing methods for verification and adapter synthesis.

- *Implementation of Advanced Event Handling:* To demonstrate the feasibility of our model, we present a prototypical implementation of our event subscription mechanism in Camunda,[1] an open-source process engine.

The remainder of this paper is structured as follows. The next section presents a scenario to motivate explicit event subscriptions in business process implementation and reviews event handling semantics in BPMN. Section 3 introduces our model for event handling. Section 4 presents formal semantics for our model and discusses how this formalisation enables the analysis of interactions. The prototypical implementation of our subscription mechanism is described in Sect. 5. Finally, Sect. 6 elaborates on related work, before Sect. 7 gives concluding remarks and outlines future work.

2   Event Subscription in Business Processes

The role of events in process implementation is illustrated in Fig. 1. Since process implementation is model-driven, first and foremost, a process is modelled visually [14]. This includes the specification of how a process is supposed to interact with its environment by subscribing to event sources and reacting to events emitted by these sources. At run-time, a process engine manages the execution of instances of the process. The detection of events that are relevant for process execution is often conducted with a separate system [8,10], e.g., a Complex Event Processing (CEP) engine [11]. CEP engines abstract the complexity of connecting to different event sources, parsing events in different formats and aggregating simple events from multiple event streams from the process recipients. Based on a subscription for events of a particular type, a CEP engine notifies the process engine about the occurrence of respective higher-level events needed for process execution.

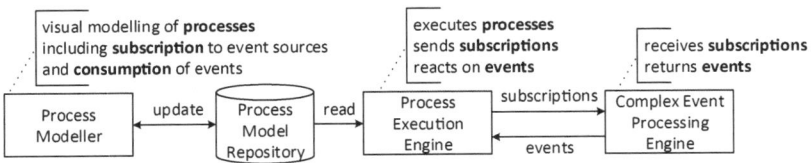

Fig. 1. Events in business process implementation.

Taking the above setting as a starting point, we first present a logistics scenario to motivate flexible handling of event subscriptions. Based thereon, we review event handling semantics as defined by BPMN and detail requirements for a more flexible model.

---

[1] https://camunda.org/.

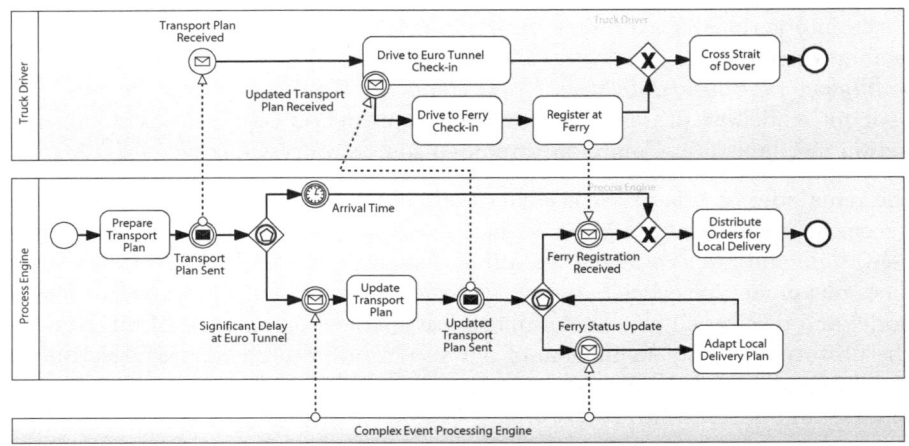

Fig. 2. Motivating scenario: A shipment process.

Motivating Scenario. We consider a use case from the logistic domain, describing the process of shipping goods by truck from the UK to continental Europe.[2] In this process, a truck needs to cross the Strait of Dover, which may be done using a train through the Euro Tunnel or by ferry. The former is the default option, unless an accident or technical failure at the Euro Tunnel render the ferry the preferred transportation option.

A model of the process is depicted in Fig. 2. In essence, a logistic company runs a process engine that coordinates the activities of the shipment process. It first prepares a transport plan, which is sent to a truck driver. The driver heads toward the check-in point at the Euro Tunnel. Yet, the environment may signal that significant delays are observed at the Euro Tunnel check-in. In this case, the process engine updates the transport plan and informs the driver to divert to the ferry.

As long as the driver has not registered on the ferry, however, status updates by the ferry operator may require further changes to the local delivery plan. Eventually, the process engine distributes the orders among the local transport partners, while the truck driver crosses the Strait of Dover.

In this scenario, the environment influencing the conduct of the shipment process may be given as a CEP engine. This engine connects to event sources such as the on-board GPS sensor of the truck, public APIs for traffic flow information,[3] the Euro Tunnel RSS feed,[4] and notifications from the ferry operator. Based thereon, the CEP engine notifies the process engine about events that signal delays at the Euro Tunnel or changes in the schedule of the ferry that may be reached after diversion of the route.

---

[2] See also the GET Service project: http://getservice-project.eu.

[3] http://webtris.highwaysengland.co.uk.

[4] http://www.eurotunnelfreight.com/uk/contact-us/travel-information/.

Event Handling Semantics in BPMN. We next consider the semantics defined by BPMN for event handling as modelled in Fig. 2. For intermediate catching events, BPMN specification [29] states:

> *'For Intermediate Events, the handling consists of waiting for the Event to occur. Waiting starts when the Intermediate Event is reached. Once the Event occurs, it is consumed. Sequence Flows leaving the Event are followed as usual'* [29] (Sect. 13.4.2).

That is, when the control-flow reaches the event construct, it is enabled and a process instance waits for the event to happen. Once it happens, the control-flow is passed to downstream activities. As such, a process instance may not react to an event that occurred *before* its control-flow reached the respective event construct.

The above semantics are a severe limitation and preclude accurate modelling of our motivating scenario. First, events that signal delays at the Euro Tunnel check-in are published by the environment at regular intervals. Since an instance of the shipment process waits for respective events only once the transport plan has been sent, a relevant event that would have led to route diversion may have been missed. Second, events on the ferry status are not published at regular intervals, but solely upon operational changes with respect to the last notification. Again, as per event handling semantics in BPMN, a process instance may miss the relevant event, since the creation of events by the environment is decoupled from the state of process execution.

Requirements. Based on the above use case and the shortcomings of current languages, we derive the following requirements for our event handling model.

R1-Flexible Event Subscription: To accurately capture the above scenario, a more flexible definition of event handling is needed. Since event production by the environment happens independent of the state of process execution, there is a need for flexibility regarding when to subscribe to an event source. In our example, subscription to delay events may be needed right from the start of an instance of the shipment process.

R2-Efficient Event Buffering: Assuming that subscription may happen before a process instance is ready to consume an event, however, multiple events may match the respective subscription, e.g., multiple delay events may have occurred after the start of a process instance, but before the transport plan has been sent. In that case, it needs to be clarified which of these events are kept by a process engine and, thus, can possibly be consumed by a process instance, and which event is eventually consumed.

3   Event Handling Model

Addressing the requirements identified above, this section proposes a new event handling model for the specification of business processes. This model is grounded in explicit event subscriptions and event buffering mechanisms. Although these concepts are generic and applicable to various process specification languages, this section also outlines the concrete realisation of this model in BPMN, using its extension mechanism.

Below, we introduce the notion of an early subscription to an event source (Sect. 3.1) to explicitly control at which point in time events start to become relevant for a process. In particular, events that occurred before an instance of a process reaches the state of being ready to consume the event may be considered (addressing R1). Next, we turn to a mechanism to retrieve events for a process instance (Sect. 3.2). While relevant events are buffered after their occurrence, policies limit the scope of events to store and define the semantics of retrieving event from buffer (addressing R2).

3.1 Early Subscription

Reflecting on the order of the subscription to event sources, the actual occurrence of an event, and its consumption by a process instance, we note that existing models define only a partial order, see [5]. That is, an event can only be consumed if both, a subscription has been issued earlier and the event actually occurred already. These temporal dependencies are illustrated by solid arrows in Fig. 3. However, models such as the one presented in [5] do not define any temporal dependency between subscription and event occurrence. In this work, we argue that such a temporal order is needed (visualised by the dotted arrow in Fig. 3) in order to obtain a model with well-defined semantics. Against this background, our model includes the notion of a *subscription task*. Such a task is a regular activity in the specification of a process, yet has specific semantics: when executed, it issues a subscription to an event source.

BPMN Extension for Subscription Tasks. Subscription tasks can be incorporated in BPMN using its dedicated extension mechanism. Inspired by recent work [7,31,35], we formalised the extension as a BPMN+X model [33]. BPMN+X is a UML Profile [28] that enables convenient specification of extension by means

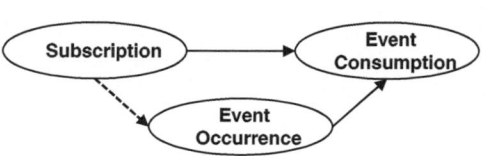

Fig. 3. Dependencies between the subscription to event sources, event occurrence, and consumption.

of stereotypes, while still being grounded in the BPMN concepts of *Extension-Definition* and *ExtensionAttributeDefinition*, which provide a standardised way of extending the language.

Specifically, the concept of a *SubscriptionTask* extends the concept of a BPMN *ServiceTask*. The rationale behind this decision is that issuing a subscription requires automatic communication with an external CEP engine. The BPMN+X model for the respective extension is given in Fig. 4, with the extended elements highlighted in grey. While *SubscriptionTask* inherits the attributes and

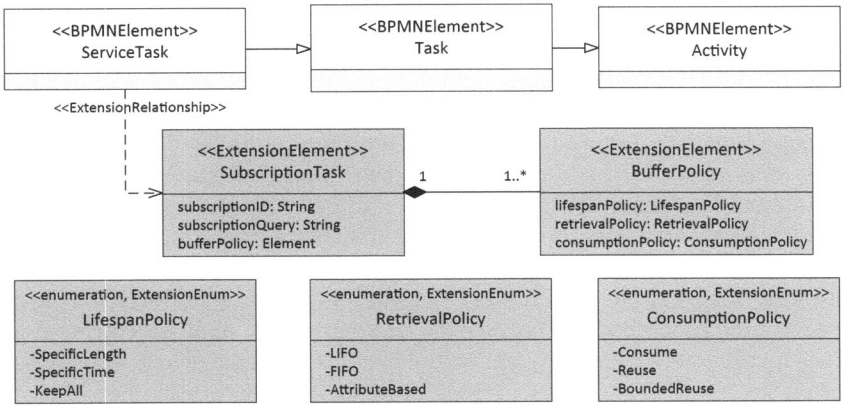

Fig. 4. BPMN+X model that captures the BPMN extension of a subscription task.

model associations of *ServiceTask*, it also has additional attributes. Attribute `subscriptionId` refers to the unique subscription identifier, needed to correlate the task with the correct event. Furthermore, an attribute `subscriptionQuery` captures the actual subscription that is registered at a CEP engine. A *Subscription Task* is further composed of another extension element, called *Buffer-Policy*, which contains three attributes (`lifespanPolicy`, `retrievalPolicy`, and `consumptionPolicy`) and will be explained below. Following the approach of [33], using the BPMN+X model, we generate the XML Schema conforming to BPMN's *ExtensionDefinition* and *ExtensionAttributeDefinition*.

The Point of Subscription. Subscription can be done at any point in time, after all the required information needed for subscription are known. If subscription is independent of instance data, then it can be done even before instantiation. Depending on domain knowledge or by analysing event logs from past execution, the earliest point to start listening for an event can be determined. If subscriptions are not done by an explicit subscription task, we resort to the standard semantics of BPMN, i.e., subscription is implicit when the control-flow reaches the respective construct. Also, for the special case of process instantiation, explicit subscription by a task is not applicable. Rather, the respective subscription to an event source is issued at process deployment.

3.2 Event Buffering

Once the subscription has been issued, whenever the event occurs, the CEP engine sends the respective event to the process engine. As an instance of the process might not yet be ready to consume the event, it is temporarily stored in a buffer. At some point in time, a process instance is ready to consume the event and thus checks, if a respective event exists in the buffer. If not, then the process waits for the event to occur. If the buffer contains a respective event, the process instance retrieves it. The event payload is then available within the context of the process instance, which continues execution.

When retrieving events from a buffer, their occurrence cardinality has to be considered. Some events may occur only once, e.g., customs clearance of a shipment at a particular border crossing will typically only be done once and thus may be represented by a single event. However, some events occur continuously, in a streaming manner. Periodical weather updates or traffic flow information for a certain route are examples of such events that are continuously emitted by the environment of a process.

If more than one event of interest (i.e., in relation to a subscription) is available in a buffer, it needs to be decided, which event shall be retrieved for a process instance. To this end, buffer policies are selected when issuing a subscription for a particular event source. These policies determine how many and which events have to be kept in the buffer and which event is selected to be consumed by a process instance. Below, we provide a basic collection of buffer policies for our event handling model.

Lifespan Policy. The lifespan policy specifies the subset of events provided by the CEP engine that shall be stored after a subscription has been issued. As such, the lifespan policy can be interpreted as determining the buffer size. Specific configuration values for this policy include:

- **Specific Length.** One may choose to store a specific number of events in the buffer. For example, only the last five events that signal the traffic flow on a particular route may be considered as relevant, as those are sufficient to establish the current trend.
- **Specific Time.** The subset of events can be selected using a time-window. Assuming that events about road incidents are emitted only once incidents actually happen, storing a specific number of these events is not useful. Rather, only events that occurred within the last 30 min may be considered to be relevant.
- **Keep All.** This configuration does not impose any restriction on storing the events. All events received after the subscription has been issued are stored in the buffer.

Retrieval Policy. The retrieval policy determines which of the stored events is most relevant for the process instance.

- **Last-In-First-Out.** There can be situations, in which the latest event supersedes previous ones and becomes the most relevant. Thinking about processes that are concerned with buying or selling of goods as part of an auction, the latest published price will be most relevant for decision-making.
- **First-In-First-Out.** Unlike the above case, the first occurrence of the event may have the highest relevance, i.e., the first event is selected for retrieval. For example, in logistics, the first operator to react to a request for last-mile delivery may be chosen for a shipment.

- **Attribute-based.** Instead of temporal aspects, the payload of an event, i.e., its attribute values may govern which event to retrieve. In the case of the above logistics example, responses to requests for last-mile delivery may indicate an expected delivery time. In that case, the operator that guarantees immediate delivery may be chosen, which is determined based on the payload of the respective event.

Consumption Policy. The consumption policy determines whether to delete an event after it has been retrieved by a process instance. Such consumption policies are well-known to influence the semantics of processing, see [15]. As part of our event handling model, we consider the following configurations regarding event consumption.

- **Consume.** According to this configuration, an event is deleted from the buffer as soon as it is retrieved by a process instance.
- **Reuse.** Here, an event can be retrieved by more than one process instance. It is worth to mention that these process instances can be of the same process as well as of different processes that run concurrently within a process engine.
- **Bounded Reuse.** In this case, a restriction is added to specify the number of times an event can be retrieved.

Application to the Motivating Scenario. Our example shipment process (Fig. 2) receives two types of events from the CEP engine. The event `Significant Delay at Euro Tunnel` becomes relevant as soon as the process is instantiated. Therefore, a subscription task is placed right after the instantiating start event construct. In the respective scenario, it takes a truck a few hours to reach the Euro Tunnel check-in. During that time, the load at the Euro Tunnel check-in can change and a delay at the check-in may influence the shipment process. Euro Tunnel publishes delay information every half an hour in their RSS feed. Thus, in our scenario, the lifespan policy is set to the last three events, so that the information about last one hour before the truck reaches the check-in is captured. The logistic company updates the transport plan based on the latest information and sets the retrieval policy as last-in-first-out. Furthermore, the event is consumed, as the information quickly becomes outdated for other shipments.

`Ferry Status Update` events become relevant solely when significant delays are observed at the Euro Tunnel check-in, since only in that case the truck is diverted. The subscription task is thus placed before the `Update Transport Plan` task. As ferry updates are not published in regular intervals, the lifespan policy is set as to a specific time in order to be informed about the current status. The retrieval policy is again last-in-first-out, as the local delivery plan is updated based on the latest information. Furthermore, events are consumed for the same reason as before.

4   Formal Semantics and Analysis

This section turns to formal analysis of business processes that communicate with their environment by means of event handling. To this end, we first present formal execution semantics for the event handling model introduced above as a BPMN extension (Sect. 4.1). In a second step, we review common analysis problems in the context of a process' interactions and outline how existing methods for verification and controller synthesis can be exploited to solve them (Sect. 4.2).

4.1   Formal Execution Semantics

Model. We define the semantics of our event handling using Coloured Petri-Nets (CPNs), a rich formalism that extends common Petri-nets with concepts for data handling. Our choice for this formalism is motivated as follows: Petri-nets have been widely used for modelling business processes [2] and many process specification languages such as BPMN and WS-BPEL can be mapped to Petri-nets to a large extent [13,23]. Moreover, Petri-nets are particularly suited to capture the interactions of a process with its environment, as they support the composition of models, following the principles of loose coupling, and assuming asynchronous communication between the components.

Next, we will give some intuitive explanations of Petri-nets and CPNs. For a comprehensive definition, we refer the reader to [32]. In essence, a Petri-net is directed graph consisting of *places* (depicted as circles) and *transitions* (depicted as squares), jointly referred to as *nodes*, that are connected by *flow* arcs. Flows can connect solely places to transitions, and vice versa. The pre-set (post-set) of a node are all nodes from which (to which) there is a flow arc to (from) the node. The state of a process is captured by a *marking*, which is a distribution of *tokens* over places. The execution semantics of a Petri-net is given by the firing rule of transitions: a transition is *enabled* in a marking, if all places in its pre-set carry at least one token. An enabled transition can fire in a marking, which changes the marking by reducing the number of tokens of places in the pre-set by one, and increasing the number of tokens of places in the post-set by one.

Coloured Petri-Nets (CPNs) extend Petri-nets with concepts for data handling. That is, places are typed with a *colourset*, and tokens are *coloured*, i.e., carry data according to the type definition of the respective place. Flow arcs are assigned *arc inscriptions*. Such inscriptions contain functions and variables, the latter are bound to the data carried by tokens. Furthermore, transitions are assigned *guard conditions*, Boolean predicates that may reference the variables of inscriptions of incoming flow arcs. Enabling of transitions is then determined for a marking and a specific binding of token data to the variables of arc inscriptions: if the guard of a transition evaluates to true under such a binding, the transition is enabled. Upon firing the transition, the inscriptions on incoming and outgoing flow arcs determine how the marking is updated, see [32].

In arc inscriptions and guards, we denote a list of m elements as $l = \langle x_1 \ldots x_m \rangle$, write $|l| = m$ for its length and refer to the i-th element as $l(i) = x_i$.

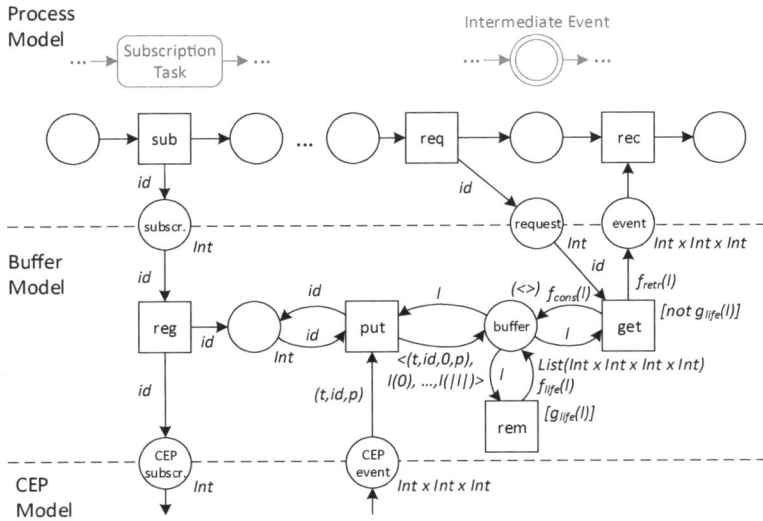

Fig. 5. CPN defining the execution semantics of early subscription and event buffering.

Semantics of Early Subscription and Event Buffering. We define the semantics of our event model using the CPN shown in Fig. 5, which is parametrised by functions used in the arc inscriptions to support the discussed policies for event buffering. In essence, this net illustrates the interplay of three components: a model of the business process, a model of the event buffer, and a model of the CEP engine. The latter is captured only by its interface that is given by two places, one capturing subscriptions sent to the CEP engine and one modelling the events sent to the buffer.

Starting with the description of the process, the step of subscribing to an event source (explicitly using a subscription task or implicitly upon reaching an event handling construct with the control-flow) is captured by a dedicated transition (sub) producing a token with the subscription identifier (on place subscr.). Here, we abstract from the subscription query. In the buffer, this token is forwarded to the CEP engine, but also kept internally to correlate the tokens that represent events produced by the CEP engine. Such tokens have a product colourset, representing the event's timestamp, identifier, and payload, here all modelled as integers. Transition put in the buffer registers events in the buffer, for which a subscription has been issued earlier. Specifically, events are kept in place buffer, which has a list colourset over the colourset of events (extended with an additional integer representing a consumption counter) and is initially marked with a token carrying an empty list (denoted by (<>)). Upon firing, transition put adds the data of the token consumed from place CEP event to this list. Which events shall be kept in the buffer is controlled by transition rem. It may fire if the transition guard g_{life} evaluates to true. If fired, it applies function $f_{\mathtt{life}}$ to the list of the token in place $\mathtt{buffer}$. This models the above lifespan policy.

Consumption of an event by a process instance is modelled as a two step-process: the process instance first requests an event (transition $\mathtt{req}$) from the buffer, before it actually receives it (transition $\mathtt{rec}$). This formalisation encapsulates the policies for event buffering in the buffer model. Specifically, transition $\mathtt{get}$ extracts an event (element in the list carried by the singleton token in place $\mathtt{buffer}$). This transition is enabled only if the transition to implement the lifespan policy is disabled. Also, the arc inscriptions $f_{\mathtt{cons}}$ and $f_{\mathtt{retr}}$ model the consumption policy and the retrieval policy, respectively.

Next, we turn to the representation of event buffering policies by instantiating the respective guards and functions.

Lifespan Policy. This policy relates to the guard condition $g_{\mathtt{life}}$ and function $f_{\mathtt{life}}$.

- *Specific Length:* Assuming that at most k events shall be kept in the buffer, the guard condition for the transition to remove events checks for the length of the respective list, i.e., $g_{\mathtt{life}}$ is defined as $|l| \geq k$. Then, function $f_{\mathtt{life}}$ selects only the k most recent events, i.e., $f_{\mathtt{life}} (l) \mapsto \langle l(n-k+1), \ldots, l(|l|) \rangle$.
- *Specific Time:* Assuming that there is global variable g in the CPN model that indicates the current time and a time window of k time units, the guard $g_{\mathtt{life}}$ checks whether some event fell out of the window, $l(i) = (t, id, n, p)$ with $t < g - k$ for some $0 \leq i \leq |l|$. The respective events are removed, i.e., $f_{\mathtt{life}} (l) \mapsto l'$ where $l'(j) = l(i) = (t, id, n, p)$ if $t \geq g - k$ and for $i - j$ events $l(m) = (t', id', n', p')$, $m < i$ it holds that $t' < g - k$.
- *Keep All:* Trivially, the guard $g_{\mathtt{life}}$ is set to false, so that function $f_{\mathtt{life}}$ does not have to be defined.

Retrieval Policy. The retrieval policy is implemented by function $f_{\mathtt{retr}}$ as follows:

- *Last-In-First-Out:* The last event of the list is retrieved, $f_{\mathtt{retr}} (l) \mapsto l(|l|)$.
- *First-In-First-Out:* This the head of the list of events is retrieved, $f_{\mathtt{retr}} (l) \mapsto l(1)$.
- *Attribute-based:* With π as a selection predicate evaluated over the payload of events, the first of events that satisfies the predicate is retrieved, i.e., $f_{\mathtt{retr}} (l) \mapsto (t, id, n, p)$, with $l(i) = (t, id, n, p)$, such that $\pi(p)$ holds true and for all $l(j) = (t', id', n', p')$, $j < i$, $\pi(p')$ is not satisfied.

Consumption Policy. Function $f_{\mathtt{cons}}$ to realise the consumption policy is defined as:

- *Consume:* The event retrieved from the buffer, assuming its position in the list of events l is i, is consumed, i.e., not written back to the buffer. This is captured by the following definition of the function implementing the consumption policy: $f_{\mathtt{cons}} (l) \mapsto \langle l(1), \ldots l(i-1), l(i+1), \ldots, l(|l|) \rangle$.

- *Reuse:* The event is not removed from the buffer, i.e., $\mathtt{f_{cons}}\,(l) \mapsto l$.
- *Bounded Reuse:* Assuming that an event can be consumed k-times and with $l(i) = (t, id, n, p)$ being the retrieved events, the function to implement the consumption policy is defined as: $\mathtt{f_{cons}}\,(l) \mapsto \langle l(1), \ldots, l(i-1), l(i+1), \ldots, l(|l|)\rangle$, if $n \geq k$, and $\mathtt{f_{cons}}\,(l) \mapsto \langle l(1), \ldots, l(i-1), (t, id, n+1, p), l(i+1), \ldots, l(|l|)\rangle$, otherwise.

4.2 Analysis Problems and Techniques

The above formalisation enables comprehensive analysis of the interactions of a business process with its environment by means of events. Petri net-based analysis of business process models has been targeted in a plethora of works [13]. These techniques become applicable in our context, based on existing formalisations of CEP engines using CPNs [26] and an unfolding of CPNs into common Petri nets [22]. Yet, we note that the separation of models for the process, the buffer, and the CEP engine provides a particularly interesting angle for analysis.

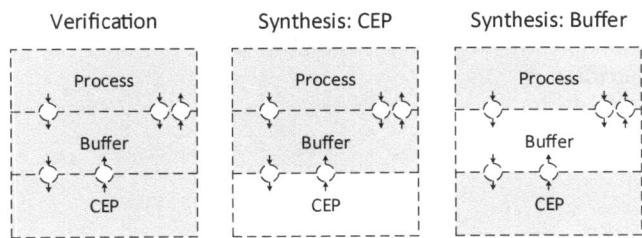

Fig. 6. Different analysis problems.

Clearly, the central goal of analysis is to ensure *correctness* of processing, which primarily relates to the interplay between the process, buffer, and CEP engine. Yet, as outlined in Fig. 6, correctness may be considered for only a subset of the models representing particular components being available. That is, traditional verification is the case of being given models of the process, the buffer, and the CEP engine. In the absence of one of these models (not having a process model is not reasonable in process implementation, though), however, existing techniques for model synthesis can also be exploited for analysis, as detailed in the remainder of this section.

Verification. If models for all components are available, the composed system can be verified for correctness. Since a CEP engine can be assumed, in general, to produce more events than can be consumed by a process instance, correctness criteria developed for distributed, interacting systems are particularly suited to be applied. A specific example for such a criterion is weak termination [1], which requires all deadlocks of the composed system to be dedicated final markings. Verification is then based on standard reachability analysis of Petri nets [32].

Synthesis. In practice, one may have solely an incomplete model of the CEP engine (e.g., a partial formalisation of event queries based on [26]) or the buffer (e.g., based on the CPN introduced above, but without an instantiation of buffering policies). In that case, the analysis problem is no longer one of verification. Rather, it can be first be approached as a problem of *controllability* [24], answering the question whether there exists at least one model that completes the composed model (i.e., completes the model of the CEP engine or of the buffer), such that the composition is correct. Once controllability has been established, adapter synthesis can be applied to generate a model for the unknown component, or part thereof, such that the composition is correct [17]. The synthesised models can be used as blueprints for the completion of the respective models, or to exemplify specific desirable situations.

5   Prototypical Implementation

To demonstrate the feasibility of our model for event handling, we developed a proof-of-concept implementation based on the open-source process engine Camunda (see Footnote 1). In this section, we first review the architecture of the prototype, before turning the to communication between the process engine and a CEP platform.

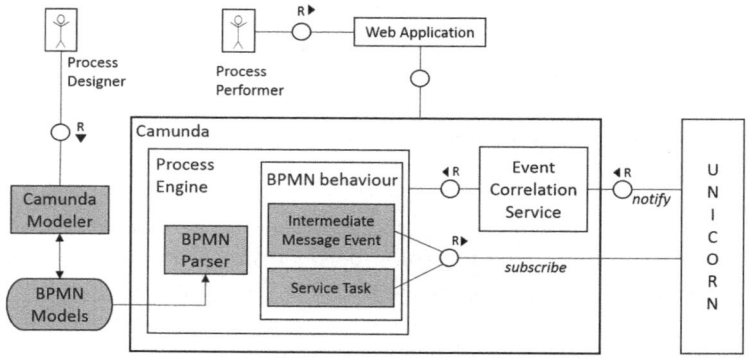

Fig. 7. Architecture of the proof-of-concept implementation in Camunda.

The overall architecture of our prototype is shown in Fig. 7, where components highlighted in grey have been extended to realise event handling according to our model. The Camunda process engine comes with a modeller, based on BPMN.io,[5] to create BPMN process models. The process models are deployed in the engine for execution. Each node in the process model has a specific *BPMN behaviour* associated to it, which determines what is executed when the node

---

[5] http://bpmn.io/.

is activated by control-flow. Camunda further provides a REST interface that enables consumption of events. Yet, it lacks support for subscriptions to external event sources. Therefore, we connected Camunda to an open-source CEP engine, the Unicorn platform.[6]

In our prototype, we supported explicit subscriptions in two ways. First, the *BPMN behaviour* of intermediate message event constructs is extended to issue subscriptions to Unicorn when activated by control-flow, which is the traditional event handling semantics in BPMN. In addition, we extended the behaviour of a service task to issue subscriptions, which enables explicit subscriptions at any point in the process model. These extensions are also reflected in the Camunda process modeller. For example, the service task is extended with attributes such as the subscription query (as detailed in the BPMN+X model in Fig. 4).

Next, we turn to the communication between the Camunda process engine and Unicorn. After the model has been deployed and the control-flow reaches the element that issues a subscription, the `execute` method of the respective element sends a POST request to Unicorn. This request includes the subscription query, the identifier of the element that sent the request, and the interface of Camunda to receive notifications.

As detailed in Fig. 8, Unicorn responds to each subscription with an Universally Unique Identifier (UUID), which is kept by the process engine for correlation purposes. When Unicorn detects an event, it sends a notification together with the UUID to Camunda, using the REST endpoint provided in the subscription as the interface. The UUID is then used by the built-in *Event Correlation Service* of Camunda to correlate each received event to the correct process instance. This is also the component responsible for evaluating the event buffering policies. A process instance that waits for an event with the respective UUID then retrieves the event and continues processing.

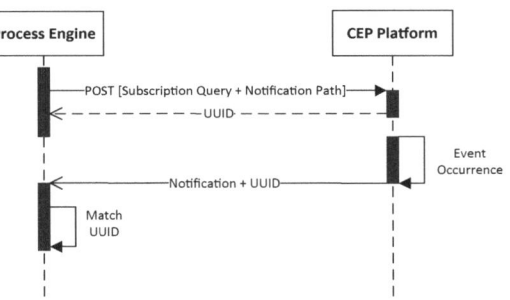

Fig. 8. Event subscription sequence.

Our proof-of-concept implementation shows that an off-the-shelf process engine for BPMN, can be extended with advanced event handling that includes early subscription and event buffering. By relying on a CEP platform to receive events, only minor changes have to be incorporated in the process engine.

---

[6] https://bpt.hpi.uni-potsdam.de/UNICORN/WebHome.

6 Related Work

The integration of event processing and business processes is an emerging field that takes the advantage of a wide variety of event sources to improve the operation of business processes [25]. In particular, events enable comprehensive monitoring of the status of a business process, thereby enabling immediate reaction to unexpected situations [7,8]. As event sources produce events in diverse formats and one event source can be relevant for several business processes, it is efficient to setup an event processing engine when exploiting events in the implementation of a business process [19]. However, it has been acknowledged that event subscription, event binding and event correlation impose severe challenges, when striving for an end-to-end integration of a event processing engine and a process engine [9]. In this context, our work proposes an event handling model that extends the conventional subscription mechanism suggested by BPMN [29] and enables explicit control of the point of event subscription.

Event subscription is a well-known area of research in the field of event processing. Publish-Subscribe systems build the interconnection between the information provider and information consumer based on a subscription-notification paradigm [20]. Traditional Pub-Sub systems manage events that occur once an explicit subscription to some event source has been issued. However, advanced systems such as PADRES [21] also provide the option to access historic as well as future data (events), and support event processing operations on them. Other proposals in this domain include content-based subscriptions systems [3]. Unlike topic/channel-based subscriptions, they give users the flexibility to subscribe to events based on the information carried by the events, i.e., based on their payload. Also, there are several middleware systems designed for event-based communication in a large-scale distributed setup [27,30]. While the above works inspired our event handling model, their focus is not the processes that consume events.

From a BPM perspective, the CASU framework [12] discusses about the subscription of events and the duration of subscriptions. Yet, it is limited to events that instantiates processes and does not include explicit subscriptions once a process instance started execution. Also, we note that enterprise integration patterns (EIP) that capture common middleware scenarios have been formalised using CPNs [16]. While this resembles our approach of defining execution semantics, the EIP formalised in [16] do not cover some aspects that are relevant from a process perspective, e.g., explicit retrieval policies.

7 Conclusion and Future Work

In this work, we took the limitations of common process specification languages when expressing complex event handling semantics as a starting point. Languages such as BPMN, UMN Activity diagrams, or WS-BPEL, do not offer flexible means to describe when to subscribe to an event source and how to retrieve an event for a process instance. The need for advanced event handling has further been motivated with a scenario from the domain of logistics.

Against this background, we presented an event handling model that is grounded in early subscriptions to event sources and event buffering. Specifically, we showed how this model is realised in BPMN. We used the BPMN extension mechanism to introduce a service task to perform early subscription and defined policies that govern how events are managed in and retrieved from a buffer. For the presented event handling model, we also presented a formal definition of execution semantics. Based thereon, we discussed analysis problems for the interactions of a process with its environment and outlined how existing reasoning techniques can be applied to solve them. Finally, we presented a prototypical implementation of our model in an open-source process engine, which highlights the feasibility of our approach. As such, we presented an end-to-end solution that covers the diverse aspects involved in event handling, from language design, through formal analysis, to technical considerations.

Our work opens several directions for future research. First and foremost, the interplay of process instances when retrieving events from a buffer deserves further investigations. For example, an event can be shared by all or a subset of instances of a single process as well as instances of multiple processes. Such scenarios may require even more fine-granular control on how events are retrieved. The next logical step will be to make unsubscription flexible as well. This is important to avoid receiving a residual event while executing a process containing loops. A systematic exploration of dependencies between buffer policies and their impact on the conduct of a process is another direction of future work.

Acknowledgements. We are grateful for comments provided by Jan Sürmeli in the course of this work. The work is partially funded by the German Research Foundation (DFG) under grant agreement number WE 4891/1-1.

References

1. van der Aalst, W.M.P., Lohmann, N., Massuthe, P., Stahl, C., Wolf, K.: Multiparty contracts: agreeing and implementing interorganizational processes. Comput. J. **53**(1), 90–106 (2010)
2. van der Aalst, W.M.P., Stahl, C.: Modeling Business Processes - A Petri net-Oriented Approach. Cooperative Information Systems Series. MIT Press, Cambridge (2011)
3. Aguilera, M.K., Strom, R.E., Sturman, D.C., Astley, M., Chandra, T.D.: Matching events in a content-based subscription system. In: PODC, pp. 53–61. ACM (1999)
4. Alves, A., et al.: Web Services Business Process Execution Language Version 2.0. Oasis standard, OASIS (2007)
5. Barros, A., Decker, G., Grosskopf, A.: Complex events in business processes. In: Abramowicz, W. (ed.) BIS 2007. LNCS, vol. 4439, pp. 29–40. Springer, Heidelberg (2007). doi:10.1007/978-3-540-72035-5_3
6. Barros, A., Decker, G., Dumas, M., Weber, F.: Correlation patterns in service-oriented architectures. In: Dwyer, M.B., Lopes, A. (eds.) FASE 2007. LNCS, vol. 4422, pp. 245–259. Springer, Heidelberg (2007). doi:10.1007/978-3-540-71289-3_20
7. Baumgrass, A., Herzberg, N., Meyer, A., Weske, M.: BPMN extension for business process monitoring. In: Enterprise Modelling and Information Systems Architectures (EMISA). LNI, GI (2014)

8. Baumgraß, A., et al.: Towards a methodology for the engineering of event-driven process applications. In: Reichert, M., Reijers, H.A. (eds.) BPM 2015. LNBIP, vol. 256, pp. 501–514. Springer, Cham (2016). doi:10.1007/978-3-319-42887-1_40

9. Beyer, J., Kuhn, P., Hewelt, M., Mandal, S., Weske, M.: Unicorn meets chimera: integrating external events into case management. In: BPM Demo Session. CEUR-WS.org (2016)

10. Bülow, S., et al.: Monitoring of business processes with complex event processing. In: Lohmann, N., Song, M., Wohed, P. (eds.) BPM 2013. LNBIP, vol. 171, pp. 277–290. Springer, Cham (2014). doi:10.1007/978-3-319-06257-0_22

11. Cugola, G., Margara, A.: Processing flows of information: from data stream to complex event processing. ACM Comput. Surv. **44**(3), 15:1–15:62 (2012)

12. Decker, G., Mendling, J.: Process instantiation. Data Knowl. Eng. **68**(9), 777–792 (2009)

13. Dijkman, R.M., Dumas, M., Ouyang, C.: Semantics and analysis of business process models in BPMN. Inf. Softw. Technol. **50**(12), 1281–1294 (2008)

14. Dumas, M., Rosa, M.L., Mendling, J., Reijers, H.A.: Fundamentals of Business Process Management. Springer, Heidelberg (2013). doi:10.1007/978-3-642-33143-5

15. Etzion, O., Niblett, P.: Event Processing in Action. Manning Publications Co., Greenwich (2010)

16. Fahland, D., Gierds, C.: Analyzing and completing middleware designs for enterprise integration using coloured Petri nets. In: Salinesi, C., Norrie, M.C., Pastor, Ó. (eds.) CAiSE 2013. LNCS, vol. 7908, pp. 400–416. Springer, Heidelberg (2013). doi:10.1007/978-3-642-38709-8_26

17. Gierds, C., Mooij, A.J., Wolf, K.: Reducing adapter synthesis to controller synthesis. IEEE Trans. Serv. Comput. **5**(1), 72–85 (2012)

18. Guabtni, A., Motahari-Nezhad, H.R., Benatallah, B.: Using graph aggregation for service interaction message correlation. In: Mouratidis, H., Rolland, C. (eds.) CAiSE 2011. LNCS, vol. 6741, pp. 642–656. Springer, Heidelberg (2011). doi:10. 1007/978-3-642-21640-4_47

19. Herzberg, N., Meyer, A., Weske, M.: An event processing platform for business process management. In: EDOC. IEEE (2013)

20. Hinze, A., Buchmann, A.P.: Principles and Applications of Distributed Event-Based Systems. Information Science Reference, Hershey (2010)

21. Jacobsen, H.A., Muthusamy, V., Li, G.: The PADRES event processing network: uniform querying of past and future events. IT - Inf. Technol. **51**(5), 250–260 (2009)

22. Jensen, K., Kristensen, L.M.: Coloured Petri nets - modelling and validation of concurrent systems. Springer, Heidelberg (2009). http://www.springer.com/gp/book/9783642002830

23. Lohmann, N.: A feature-complete Petri net semantics for WS-BPEL 2.0. In: Dumas, M., Heckel, R. (eds.) WS-FM 2007. LNCS, vol. 4937, pp. 77–91. Springer, Heidelberg (2008). doi:10.1007/978-3-540-79230-7_6

24. Lohmann, N., Wolf, K.: Realizability is controllability. In: Laneve, C., Su, J. (eds.) WS-FM 2009. LNCS, vol. 6194, pp. 110–127. Springer, Heidelberg (2010). doi:10. 1007/978-3-642-14458-5_7

25. Luckham, D.C.: The Power of Events: An Introduction to Complex Event Processing in Distributed Enterprise Systems. Addison-Wesley, Boston (2010)

26. Macià, H., Valero, V., Díaz, G., Boubeta-Puig, J., Ortiz, G.: Complex event processing modeling by prioritized colored Petri nets. IEEE Access **4**, 7425–7439 (2016)

27. Meier, R., Cahill, V.: Steam: event-based middleware for wireless ad hoc networks. In: ICDCS Workshops, pp. 639–644 (2002)

28. OMG: Unified Modeling Language (UML), Version 2.5 (2012)
29. OMG: Business Process Model and Notation (BPMN), Version 2.0, January 2011
30. Pietzuch, P.R., Bacon, J.M.: Hermes: a distributed event-based middleware architecture. In: ICDCS Workshops, pp. 611–618 (2002)
31. Pillat, R.M., Oliveira, T.C., Alencar, P.S.C., Cowan, D.D.: BPMNt: A BPMN extension for specifying software process tailoring. Inf. Softw. Technol. **57**, 95–115 (2015)
32. Reisig, W.: Understanding Petri nets - Modeling Techniques, Analysis Methods, Case Studies. Springer, Heidelberg (2013). doi:10.1007/978-3-642-33278-4
33. Stroppi, L.J.R., Chiotti, O., Villarreal, P.D.: Extending BPMN 2.0: method and tool support. In: Dijkman, R., Hofstetter, J., Koehler, J. (eds.) BPMN 2011. LNBIP, vol. 95, pp. 59–73. Springer, Heidelberg (2011). doi:10.1007/978-3-642-25160-3_5
34. Weske, M.: Business Process Management - Concepts, Languages, Architectures, 2nd edn. Springer, Heidelberg (2012). doi:10.1007/978-3-642-28616-2
35. Yousfi, A., Bauer, C., Saidi, R., Dey, A.K.: uBPMN: a BPMN extension for modeling ubiquitous business processes. Inf. Softw. Technol. **74**, 55–68 (2016)

Artifact-Driven Monitoring for Human-Centric Business Processes with Smart Devices: Assessment and Improvement

Giovanni Meroni$^{(\boxtimes)}$ and Pierluigi Plebani

Dipartimento di Elettronica, Informazione e Bioingegneria, Politecnico di Milano,
Piazza Leonardo da Vinci, 32, 20133 Milan, Italy
{giovanni.meroni,pierluigi.plebani}@polimi.it

Abstract. Monitoring human-centric business processes requires human operators to manually notify to a BPMS when activities start or end. Even if nowadays smart devices, like smartphones and tablets, are adopted to make the transmission of these notifications easier, such devices usually hold a *passive* role, being a simple mediator between the BPMS and human operators.

In this paper, we adopt the Internet of Things (IoT) paradigm by envisioning an artifact-driven process monitoring where all the objects interacting with a business process instance can be coupled with a smart device to *actively* detect when process activities start or end. To support the artifact-driven monitoring, we propose an ontology-based approach to assess and improve the monitorability of a process model.

Keywords: Human-centric processes · Business process monitoring · Artifact-driven process monitoring · Ontology · Internet of Things

1 Introduction

Especially for human-centric business processes, there could be a gap between the real execution of a process, and what has been recorded in the process execution log (the event log, hereafter) by a Business Process Management System (BPMS). As pointed out by [4], when comparing the actual execution of a business process to the notifications stored in the event log, four cases are possible: (i) an activity is performed and the related start/end notifications are correctly recorded in the event log, or (ii) the activity is not performed an no registrations are stored in the event log. In these two cases, no misalignment occurs. On the other hand, if (iii) notifications about an activity that has been really executed are not recorded in the event log, then we are in the case of *invisible events*. Similarly, (iv) if an activity is not performed but a registration is present in the event log, we are in a *false events* situation. In a process being completely and correctly monitored, only the former two cases should occur. On the other hand, the presence of invisible and false events during execution denote that the process monitoring is unreliable.

© Springer International Publishing AG 2017
J. Carmona et al. (Eds.): BPM Forum 2017, LNBIP 297, pp. 160–176, 2017.
DOI: 10.1007/978-3-319-65015-9_10

When a business process is completely automated, invisible and false events rarely occur, as the BPMS directly controls when activities should be executed. In case of human-centric business processes, on the other hand, invisible and false events become more likely to occur. Indeed, common BPMSs usually propose just a list of activities to the operator responsible for their execution. It is then up to the operator notifying the BPMS when each activity is initiated and when it completes. However, this task requires the user to stop its own duties to interact with the BPMS. Therefore, it is prone to being forgotten, delayed, or erratically performed, either accidentally or intentionally.

To reduce the probability of such misalignments, we adopt an *artifact-driven monitoring* solution [2]. Artifacts are a generalization of the physical objects participating to a process and when the process is running, they are instantiated by actual physical objects. Instead of focusing only on monitoring the control flow, we also monitor the state of the physical artifacts manipulated during the execution of the business process. In particular, we use the existing relationships – expressed in the business process model – between the artifacts and the activities that determine a change in the state of such artifacts.

Goal of this work is to exploit the Internet of Things (IoT) paradigm to couple physical objects instantiating the artifacts with smart devices (i.e., single board computing devices, like Raspberry PI or Intel Galileo, equipped with sensors and communication interfaces), so that they could became *smart objects* [17]. This allows us to easily automate the generation of the events when a human activity starts and completes.

As there could be a gap between the capabilities offered by a smart object, in terms of sensors, and the information needed to derive the state of the artifact instantiated by such a smart object, this paper proposes an ontology-based approach able to: (i) assess the monitorability [10] of an artifact given the smart objects instantiating such an artifact; (ii) suggest modifications to the smart objects and the process model to improve the monitorability.

This paper is structured as follows: Sect. 2 briefly describes the basics of artifact-driven monitoring and the problem statement. Section 3 introduces the ontologies used to describe the capabilities of smart objects, while the algorithms to assess the monitorability are presented in Sect. 4. Section 5 introduces possible approaches to improve both the process model and the monitoring infrastructure to increase the monitorability. Finally, Sect. 6 surveys the state of the art, and Sect. 7 concludes this paper outlining possible future work.

2   Artifact-Driven Monitoring

Process monitoring usually relies on the information stored into the event log, which contains what has been observed by a BPMS during the execution of a process (e.g., when each activity has started and/or completed its execution). Typical monitoring approaches analyze these events to estimate the performance of each process execution (e.g., assessing process/activity completion time, throughput, resources saturation, etc.) [22]. In some cases, the monitoring focuses on checking the so-called process conformance, i.e., the alignment between the process execution and the model [14].

Especially in case of human-centric business processes, it may happen that the process continues being executed even if, for a period of time, the BPMS is not informed about the on-going activities. In this case, monitoring the execution of the process at runtime may experience delays or inconsistencies. Therefore, a complete and consistent analysis can be done only post-mortem.

To solve this issue, we propose an artifact-driven monitoring platform [2], where the process is monitored by monitoring the artifacts manipulated during the process execution. Artifacts and states are represented in the process model and associated to the process activities to specify the state held by the artifacts before and after the execution of such activities. This way, by monitoring when artifacts transition to a new state, we are able to determine when the associated activities start or end.

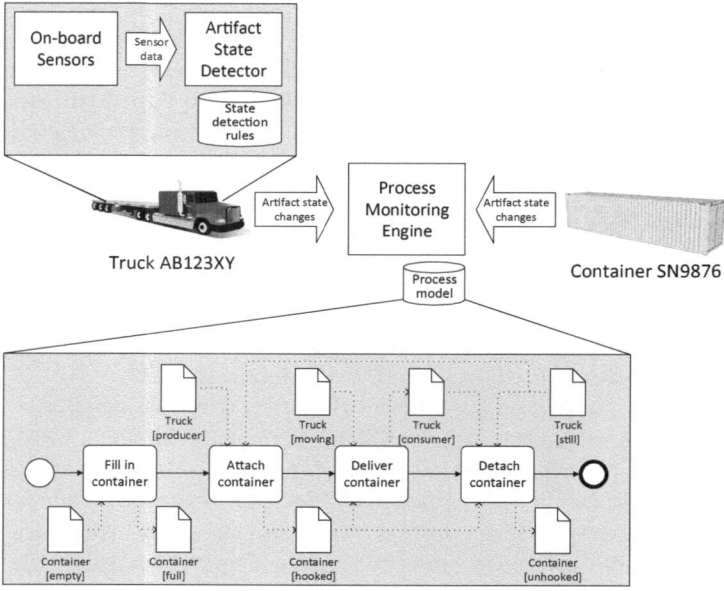

Fig. 1. Architecture of an artifact-driven monitoring platform.

On this basis, being able to capture the information about the state of an artifact becomes fundamental. To this aim, we adopt the IoT paradigm: i.e., physical objects can be turned into smart objects by equipping them with sensors, a single board computing device, and a transmission interface. Since physical artifacts are instantiated by physical objects when the process is running, such smart objects can become self-aware of their own state. Therefore, information about the state of the artifacts can be automatically collected and sent to a monitoring engine that can assess the process conformance at run-time, as long as the engine receives this information as soon as the state of the artifact changes[1] Figure 1 shows an architecture supporting the artifact-driven monitoring solution applied to a simple running example in the logistics domain[2]. The manufacturer ACME has to send some of its products to one of its customers. To do so, it firstly fills in one of its shipping *containers* by putting the products to be sent into it. Then, the container is attached to a *truck*, which delivers the container to the customer. Finally, once the truck reaches the customer, the container is detached from the truck. As both the truck and the container are physical artifacts actively participating to the process, they are instantiated by smart objects that are composed by the *On-board Sensors* and *Artifact State Detector* modules.

The *On-board Sensors* module is responsible for sensing the physical properties of the attached object (e.g., truck and container) to determine the state of such an object. For instance, if the smart object embodying the container is equipped with scales, and the scales do not detect any load, then the container can be considered as *empty*. To make this detection possible, the data provided by the On-board Sensor module feed the *Artifact State Detector* which compares those data to the *state detection rules*. These rules predicate on sensors values to determine the current state of the artifact. For instance, for the smart object embodying the container, the states to be monitored will be *empty*, *full*, *hooked* (attached to a truck) and *unhooked* (detached from a truck).

When the Artifact State Detector realizes that the artifact has changed its status, it will inform the *Process Monitoring Engine*. This way, the Process Monitoring Engine, as it communicates with all the smart objects instantiating the artifacts included in the process model, can detect violations at the process instance level (i.e., for each specific execution of the process). For example, if the container is attached to the truck before being filled in, then the container will notify to the Process Monitoring Engine that its state has changed from *empty* directly to *hooked*. Then, the engine will detect that the activity Attach container was started before the activity Fill in container was completed.

Given a process model, the quality of an artifact-driven monitoring highly depends on the capabilities of the smart objects that will instantiate the artifacts once the process instances are running. One or more smart objects may lack sensors or state detection rules needed to determine when they are in a specific state. As such, for the process instances that use these smart objects, determining when the artifacts instantiated by these smart objects assume that state would be impossible. The lack of this information negatively affects the so-called process monitorability, i.e., how many activities can be monitored with respect to the set of activities composing the process model [10].

---

[1] Considerations on how to practically deal with sensor event streams are outside the scope of this paper.

[2] For the sake of clarity, we applied our architecture to a simple scenario. However, this architecture can also monitor complex multi-party business processes [15].

To this aim, this paper proposes the usage of ontologies to describe smart objects and state detection rules. These ontologies can be exploited to: (i) describe the *monitoring infrastructure* available to monitor a given process, (ii) assess the monitorability of the process, and (iii) suggest modifications about the artifacts defined in the process model, the capabilities of the smart objects instantiating them, or the state detection rules.

It is worth noting, that the logic behind the analysis performed by the Process Monitoring Engine is not considered here as it has been already discussed in [2].

3   Ontologies

To assess the process monitorability, the presented approach relies on two ontologies. Firstly, a *Smart Object ontology* describes the capabilities of a smart object in terms of sensors and measured physical properties. Secondly, a *State Detection Rules Ontology* formalizes the dependencies between the states of an artifact and its physical properties.[3]. The advantages of adopting ontologies with respect to other data structures (i.e., databases) are interoperability, communication, and reusability [20]. As discussed in Sect. 6, many ontologies for the Internet of Things are currently being developed and populated with information describing smart devices, sensors, and their capabilities. Therefore, to describe the monitoring infrastructure, plenty of information on devices, measurements and formulas are already available and don't have to be defined from scratch. In addition, relationships among concepts belonging to different ontologies can be defined, and data can be easily integrated. Finally, the use of ontologies simplifies the integration of monitoring infrastructures belonging to different organization. This is particularly useful in multi-party processes, where organizations interact with smart objects belonging to other organizations.

In the following subsections the structure of these ontologies will be introduced. Then, the monitoring infrastructure will be defined in terms of the individuals populating the ontologies.

3.1   Smart Objects Ontology

To represent the capabilities of a smart object, we chose to adopt and extend the FIESTA-IoT ontology [1], which is one of the most comprehensive ontologies for describing smart objects. In particular, FIESTA-IoT combines the W3C SSN ontology [6] – describing the characteristics of sensors – with a subset of the M3 ontology [11] (M3-Lite) – providing a taxonomy to categorize the smart objects with respect to their function (e.g., sensor, actuator, etc.), and the sensed information. Other ontologies, such as QU[4] for standardizing the physical properties and the units of measure of the data provided by each sensor, are also combined.

---

[3] The proposed ontologies have been implemented with Protégé [18] and are available at http://purl.org/polimi/martifact/sosdr.

[4] See http://purl.org/NET/ssnx/qu/qu.

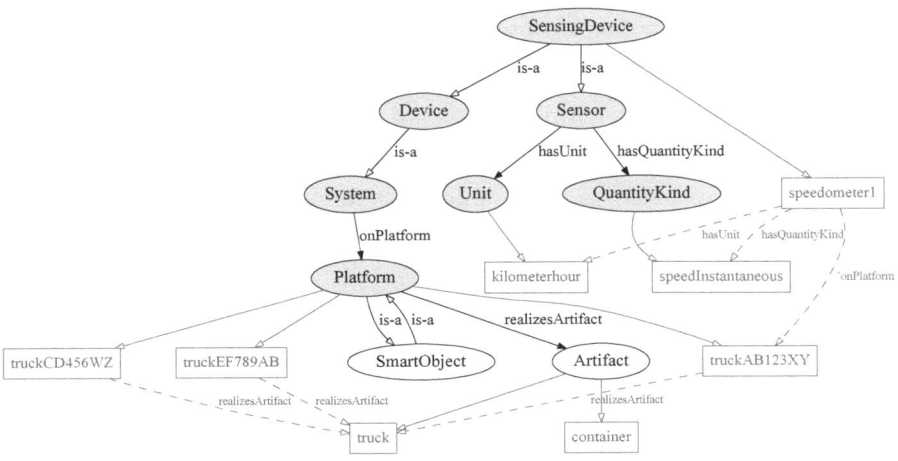

Fig. 2. Smart Objects ontology. Circles represent the classes (classes belonging to FIESTA-IoT are grayed out). Rectangles represent individuals. Dashed lines represent property assertions among individuals.

Figure 2 illustrates the (simplified) structure of the *Smart Objects ontology*, populated with some individuals.

In FIESTA-IoT, hardware devices are modeled with a hierarchy of classes: *System*, which describes a generic hardware device, *Device*, a specialization of *System* representing a hardware device dedicated to a specific purpose, and *Sensing device*, representing a hardware device dedicated to sense a physical property. *Sensing device* is also a specialization of the *Sensor* class, which represents an instrument (not necessarily electronic) to sense a physical property. In this paper, with the term *sensor*, we will always refer to a sensing device.

The *QuantityKind* class, linked to *Sensor* via the *hasQuantityKind* object property, indicates the physical property that is measured by a sensor. For example, to indicate that scales measure the weight, an individual *scales* of the *Sensor* class, an individual *weight* of the *QuantityKind* class, and an assertion of the *hasQuantityKind* object property among *scales* and *weight* have to be added to the ontology.

The *Unit* class, linked to *Sensor* via the *hasUnit* object property, indicates the unit of measure that is used by a sensor to represent a physical property. For example, to indicate that scales express the weight in kilograms, an individual *kilogram* of the *Unit* class, and an assertion of the *hasUnit* object property among *scales* and *kilogram* have to be added to the ontology. Hardware devices (i.e., *System* elements) can also be aggregated to constitute an IoT platform, which is represented by the *ssn:Platform* class. A platform can be roaming (which is represented by the *iot-lite:isMobile* data property of a platform), or can be fixed (i.e., resides on a specific location).

FIESTA-IoT does not represent the relationship between the physical objects and the abstract artifact they impersonate. Also, the concept of smart object, i.e., a physical object equipped with sensors, is not explicitly defined. Therefore, we extended FIESTA-IoT with the following concepts:

SmartObject class. This concept is equivalent to the *ssn:Platform* class. Like a platform, a smart object is made of different components (i.e. devices), which may be sensors, actuators, computational or transmission modules. For example, a truck whose license plate is AB123XY is represented as an individual *truckAB123XY* of the *SmartObject* class.

Artifact class. This concept represents the physical artifacts that can be instantiated by smart objects. For example, a generic truck is represented as an individual *truck* of the *Artifact* class.

realizesArtifact object property. This property associates the *SmartObject* concept to the *Artifact* one. This way, it is possible to describe which smart objects instantiate an artifact. For example, to specify that the truck whose license plate is AB123XY is a truck, the individual *truckAB123XY* is linked to the *truck* one by using *realizesArtifact*.

3.2 State Detection Rules Ontology

To formalize how the data coming from sensors can be used to infer the state of an artifact, we adopted and extended the Physics Domain ontology presented in [12]. The main advantage of this ontology is the possibility to define interdependencies among physical concepts. To do so, formulas to derive a physical concept given other ones (i.e., speed given space and time) are modeled in the ontology. This turns to be useful when new state detection rules have to be modeled, and it will be discussed in detail in Sect. 5.

In the Physics Domain ontology, such conversion formulas are modeled with the *Formula* class. To indicate which physical concepts are required for and derived from a conversion formula, the *Parameter* class is introduced and linked to *Formula* with, respectively, the *hasInput* and *hasOutput* object properties. To specify the physical concept of a parameter, *Parameter* is linked to the *QuantityKind* concept, imported from FIESTA-IoT, by the *hasConcept* object property. To specify the unit of measure of a parameter, *Parameter* is linked to the *Unit* concept, imported from FIESTA-IoT, by the *expressedInUnit* object property[5].

However, in the Physics Domain ontology dependencies among physical concepts and states that can be assumed by the artifacts are not defined. Therefore it is not possible to express which physical concepts are needed to determine when an artifact assumes a specific state. For this reason, in addition to the concepts presented in [12], we introduced the following elements in the State Detection Rules ontology (Fig. 3):

---

[5] Actually, the Physics Domain ontology uses the name *hasUnit* to indicate such a property. However, *hasUnit* has already been used in FIESTA-IoT to indicate the relation between *Sensor* and *Unit*. Hence, to avoid ambiguity, we changed the name.

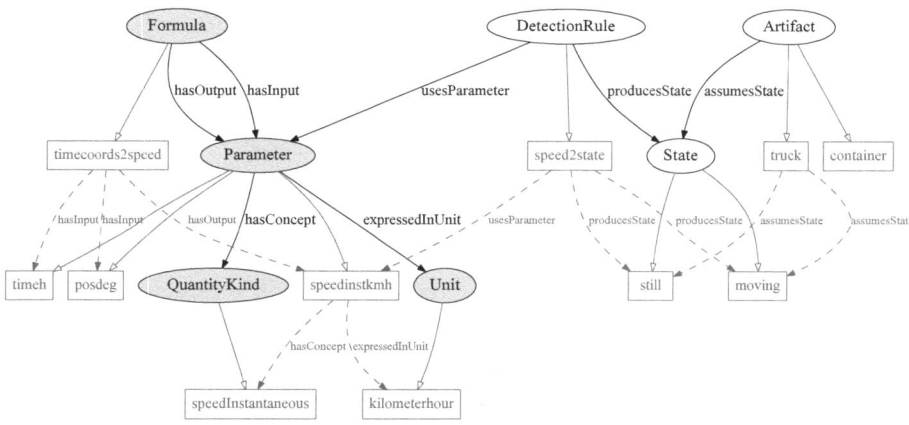

Fig. 3. State Detection Rules ontology. Circles represent the classes (classes belonging to the Physics Domain Ontology [12] are grayed out). Rectangles represent individuals. Dashed lines represent property assertions among individuals.

State class. This concept represents the discrete states that may be derived from sensor values. For example, the state *moving*, indicating that an artifact is moving, is represented as an individual of the *State* class.

assumesState object property. This property associates the *Artifact* concept, imported from the Smart Objects ontology, to the *State* one. This way, it is possible to indicate all the possible states that an artifact may assume. For example, to specify that a truck can be moving, the individual *moving* is linked to the *truck* one by using *assumesState*.

DetectionRule class. This concept represents the state detection rules. For example, the rule *speed2state*, that determines if an artifact is still or moving based on its speed expressed in kilometers per hour, is represented as an individual of the *DetectionRule* class. Note that a state detection rule is different from a conversion formula: the former derives discrete states from physical concepts, while the latter converts sets of physical concepts into other physical concepts. Therefore, the *Formula* class cannot be used for state detection rules.

usesParameter object property. This property associates the *DetectionRule* concept to the *Parameter* one. This way, it is possible to formalize which input data are required by the state detection rules. For example, to specify that the rule *speed2state* requires the speed of the artifact expressed in kilometers per hour as input parameter to operate, the individual *speedkmh*, referencing the physical concept of speed expressed in kilometers per hour, is linked to the *speed2state* one by using *usesParameter*.

producesState object property. This property associates the *DetectionRule* concept to the *State* one. This way, it is possible to formalize which state detection rule can be used to derive a state. For example, to specify that the state *moving* can be derived by using the rule *speed2state*, the individual *moving* is linked to the *speed2state* one by using *producesState*.

4   Process Monitorability with Smart Objects

Guceglioglu et al. [10] measure the monitorability of a process by comparing the number of activities whose execution status (i.e., if they are started or ended) cannot be known to the total number of activities. This work takes for granted that activities either can be fully monitored or cannot be monitored at all. Actually, in some cases an activity could be *partially* monitored: e.g., when it is possible to know when an activity starts but not when it ends (or vice-versa).

Our goal is to measure the monitorability of a process model, assuming that each execution of the same process represented in the model may differ from the other ones in terms of the capabilities and configurations of the smart objects instantiating the artifacts. For example, some containers may have the scales and some others may not, or only some trucks may have a speedometer. Thus, depending on the specific container and truck used by a process execution, the execution of some activities could not be completely monitorable. Moreover, we want to distinguish between the activation and termination conditions, to give a finer granularity in the assessment of the monitorability.

Assessing the monitorability of a process model helps the process designer to estimate the overall accuracy of the monitoring once the process is executed. Moreover, our approach can also be useful to guide the designer to improve the monitoring, as we are able to identify which activities are the most trouble-some to monitor, and how to improve their monitorability by suggesting how to reconfigure the smart devices in terms of sensors and state detection rules.

4.1   Problem Setting

To measure the process monitorability, we assume that the ontologies contain the information about available smart objects, along with their capabilities. Thus, hereafter, $I = \{T^{SO}, T^{SDR}\}$ represents the monitoring infrastructure defined by the individuals of both the Smart Objects and State Detection Rules ontologies.

Moreover, we assume that the process to be monitored is modeled using a notation, like BPMN 2.0, where couples of artifacts and states can be associated to the activities composing the process, either as input or as output. This way, the state held by the artifacts before and after the execution of an activity can be interpreted as the condition determining, respectively, the activation and termination of the activity. This way, a process model P is defined as $P = \{A_i\}$, where:

$$A_i = \langle name, C_i^{start}, C_i^{stop} \rangle$$

is an activity defined by its name, the condition determining its activation, and the condition determining its termination.

Referring to the example in Fig. 1, we have four activities:

$$A_1 = \langle Fill\,in\,container, C_1^{start}, C_1^{stop}\rangle, \quad A_2 = \langle Attach\,container, C_2^{start}, C_2^{stop}\rangle,$$

$$A_3 = \langle Deliver\,container, C_3^{start}, C_3^{stop}\rangle, \quad A_4 = \langle Detach\,container, C_4^{start}, C_4^{stop}\rangle.$$

A condition determining the activation (termination) of an activity is defined as:

$$C_i^{start} = \{ARS_{i,j}\}, \quad C_i^{stop} = \{ARS_{i,k}\}$$

where $ARS_{i,j} = ARS_{i,k} = \langle artifact, state\rangle$ is the artifact, along with the state it assumes, that in the process model is associated to the i-th activity as input (output). A condition is true when all the associated artifacts are in the specified state.

For example, the conditions C_3^{start} and C_3^{stop}, determining, respectively, the activation and the termination of activity A_3 *Deliver container*, are defined by:

$$C_3^{start} = \{\langle Container, hooked\rangle, \langle Truck, moving\rangle\}, C_3^{stop} = \{\langle Truck, consumer\rangle\}$$

Thus, only when the container is hooked and the truck is moving, then we can consider the delivery started. Similarly, when the truck is in the consumer premises the delivery has terminated.

Based on this formulation, we want to assess the monitorability of a process model P with respect to the monitoring infrastructure I. Indeed, the ability of a smart object instantiating *Container* to detect if it is *hooked* influences the monitorability of the activity *Deliver container*. If no smart object has such ability, then not only this activity, but also the *Attach container* and *Detach container* ones can never be completely monitored, thus affecting the monitorability of the whole process.

4.2 Process Model Monitorability Assessment

To assess the monitorability of a process model P, we must first consider, for each couple $ARS_{i,j}$ in the process, to which extent is the monitoring infrastructure I suited to determine if $ARS_{i,j}.artifact$ assumes $ARS_{i,j}.state$.

We name this property, which can be computed querying the ontologies, *artifact state monitorability* $Mon^{ARS}(\langle artifact, state\rangle, I)$. To do so, we firstly identify the set of smart objects (i.e., all the individuals of the SmartObject class) able to instantiate the *artifact*:

$$SSO = \{so_n \mid so_n \in I.T^{SO} \wedge so_n.realizesArtifact = artifact\}$$

Then, we identify $\overline{SSO} \subseteq SSO$ containing only those smart devices so_n whose sensors can be used to detect *state*. For each so_n, the existence of at least one detection rule dr_d to derive *state*, whose input parameters par_p can all be provided by the sensors of so_n, is verified. To do so, for every par_p belonging to dr_d, the existence of a sensor device $sensDevice_s$ belonging to so_n whose measured quantity and unit of measure are the same as par_p, is verified.

$$\exists dr_d \in I.T^{SDR} \mid dr_d.producesState = state \ \land$$

$$\forall par_p \in I.T^{SRD} \mid par_p \in dr_d.usesParameter \ \land$$

$$\exists sensDevice_s \in I.T^{SO} \mid sensDevice_s.onPlatform = so_n \ \land$$

$$par_p.hasConcept = sensDevice_s.hasQuantityKind \ \land$$

$$par_p.expressedInUnit = sensDevice_s.hasUnit \ \land$$

The cardinality of the set SSO and $\overline{SSO}$ determines the artifact state monitorability, computed as:

$$Mon^{ARS}(\langle artifact, state \rangle, I) \rightarrow [0,1] = |\overline{SSO}|/|SSO|$$

The Mon^{ARS} is the basic building block for computing the process model monitorability as we need to check, for each activity belonging to the project, to which extent its activation and termination condition can be monitored, by computing the *condition monitorability* Mon^C. More formally, given $A_i = \{name, C_i^{start}, C_i^{stop}\} \in P$:

$$Mon^C(A_i.C_i^{start}, I) \rightarrow [0,1] = \prod^{ARS_{i,j} \in A_i.C_i^{start}} Mon^{ARS}(ARS_{i,j}, I) \qquad (1)$$

$$Mon^C(A_i.C_i^{stop}, I) \rightarrow [0,1] = \prod^{ARS_{i,k} \in A_i.C_i^{stop}} Mon^{ARS}(ARS_{i,k}, I) \qquad (2)$$

Based on this definition, the monitorability of a condition $A_i.C_i^{start}$ determining the activation of A_i depends on the artifact state monitorability of the couples $ARS_{i,j}$ belonging to the condition. As all the couples $ARS_{i,j}$ are required to detect when A_i starts, their contribution is computed as the product of the monitorability of each couple $Mon^{ARS}(ARS_{i,j}, I)$. The same consideration holds for the condition $A_i.C_i^{stop}$ determining the termination of A_i.

The monitorability of an activity is then defined by the monitorability of its activation and termination conditions, so that:

$$Mon^A(A_i, I) \rightarrow [0,1] = \frac{1}{2} \cdot \left(Mon^C(A_i.C_i^{start}, I) + Mon^C(A_i.C_i^{stop}, I) \right) \qquad (3)$$

We assume that the importance of determining when an activity starts or terminates is the same. It is worth noting that, if no start (termination) condition is put in the process model for a given activity, that activity is expected to be run by a BPMS (i.e., the activity is automated). Therefore, the contribution of Mon^C is 1, as the monitoring platform would always receive automatic notifications from the BPMS.

Finally, we can define the process model monitorability as:

$$Mon^P(P, I) \to [0, 1] = \frac{\sum^{A_i \in P} Mon^A(A_i, I)}{|A_i \in P|} \qquad (4)$$

Based on this definition, the process monitorability represents the average of the activity monitorability $Mon^A(A_i, I)$ computed for all the activities $\{A_i\}$ composing P. Following the same approach as [10], at this stage we do not consider the control flow as well as the probability of taking branches. Future work will aim to deal with this issue.

5   Monitorability Improvement

In addition to quantitatively assessing the process monitorability, our ontology-based approach can also assist the designer to improve it in case it does not meet the designer's expectations (i.e., process monitorability value is too low). Indeed, by analysing the monitorability at the different levels of granularities, the designer can identify which $\langle artifact, state \rangle$ couples contribute more to lowering the monitorability value. For each of these $\langle artifact, state \rangle$ couples, different strategies can be followed: (i) improving the process model, (ii) improving the infrastructure, or (iii) improving the state detection rules.

It is worth noting that modifications done to the process model or to the infrastructure are not independent: once a single modification is introduced, the contribution of the other modifications change. Therefore, the contribution to the monitorability of a process should be computed from scratch whenever a new set of modifications is defined.

5.1   Process Model Improvement

The process model improvement strategy can be adopted if the designer realizes that a low monitorability value is caused by the impossibility, for the given infrastructure, to evaluate an $\langle artifact, state \rangle$ couple, i.e., $Mon^{ARS}(\langle artifact, state \rangle, I) = 0$. Assuming that the infrastructure cannot be modified (this case will be discussed in the next paragraph), the process designer has two possibilities: (i) find, for the same artifact, a monitorable state, i.e., $\langle artifact, state' \rangle$. (ii) Find a different artifact able to monitor the same state, i.e., $\langle artifact', state \rangle$.

In the former case, by querying the ontologies, we can obtain the alternative states $state'$, for the same artifact $artifact$ already specified in the model, for which the infrastructure can ensure a better monitorability:

$$state' \in I.T^{SDR} \ \lor \ state' \in artifact.assumesState \ \lor$$

$$Mon^{ARS}(\langle artifact, state' \rangle, I) > 0$$

For example, if $Mon^{ARS}(\langle truck, moving \rangle, I) = 0$ and $Mon^{ARS}(\langle truck, accelerating \rangle, I) > 0$, the ontologies suggest to replace $\langle truck, moving \rangle$ with $\langle truck, accelerating \rangle$.

In the latter case, the ontologies suggest to use another artifact *artifact'* for which the sensors are able to return the occurrence of the same state *state*:

$$artifact' \in I.T^{SDR} \quad \vee \quad state \in artifact'.assumesState \quad \vee$$

$$Mon^{ARS}(\langle artifact', state \rangle, I) > 0$$

For instance, if the $\langle container, hooked \rangle$ cannot be monitored, as no container has the related sensors, it could happen that the truck has these sensors. Thus, replacing that couple with $\langle truck, hooked \rangle$ would improve the monitorability.

In any case, the designer will be responsible for deciding which modifications can be applied to the process model without changing its behavior. Indeed, being a domain-dependent problem, our approach can only provide suggestions.

5.2   State Detection Rules Improvement

Another possible improvement strategy consists in the introduction of new state detection rules. In particular, given an $\langle artifact, state \rangle$ couple for which more than one smart object instantiating the *artifact* is not able to monitor *state*, a new state detection rule, whose input parameters can all be provided by the smart object, can be introduced. This is only possible when a cause-effect relation among the values coming from the sensors of the smart object and *state* exists.

In some cases, a state detection rule dr_d to infer *state* may require input parameters expressed with an unit of measurement different that the one used by the sensors of a smart object so_n. This causes so_n to be unable to monitor *state*. However, this issue can be solved by simply adding a new detection rule dr'_d identical to dr_d except for the presence of a conversion formula. For example, suppose that the speedometer of the truck *truckCD456WZ* expresses the speed in miles per hour. Yet, the state detection rule *speed2state*, to infer when the truck is *moving*, requires as input the speed expressed in kilometers per hour. This causes this truck to be unable to detect when it is moving. However, a new state detection rule can be easily derived from *speed2state* by simply converting the unit of measure of the input parameter.

By querying the ontologies in the same way as when identifying $\overline{SSO}$, except for removing the constraint on $par_p.expressedInUnit$, we can identify which detection rules should be used, and how they would impact on the monitorability.

Another case concerns a state detection rule requiring one parameter that cannot be provided by the smart objects. Yet, that parameter can be derived from the ones provided by the smart objects. For example, suppose that the truck *truckEF789AB* is not equipped with a speedometer, yet it has a GPS transponder and an internal timer. In this case, the state detection rule *speed2state* cannot be used, as it requires the speed as input. However, knowing the instantaneous position of the truck from the GPS transponder, and the current date and time from the timer, it is possible to derive the speed from these physical concepts by applying the *timecoords2speed* conversion formula. Therefore, *speed2state* and *timecoords2speed* can be used to derive a new state detection rule that uses these physical concepts as input parameters, instead of the speed.

Also in this case, the ontologies can provide insights on which detection rules and conversion formulas to use. In particular, the ontology should be queried in the same way as before, except for requiring every parameter of dr_d to be either provided by a smart object so_s, or by a conversion formula cf_f, whose input parameters are either provided by so_s or, recursively, by another conversion formula cf_f'.

5.3   Infrastructure Improvement

The infrastructure improvement strategy can be adopted when the designer detects that an $\langle artifact, state \rangle$ couple has a low monitorability value due to limited capabilities offered by the smart objects instantiating the artifact. In this case, by querying the ontology, we can obtain information on how to alter the existing smart objects. In particular, for every smart object $so_n \in SSO \backslash \overline{SSO}$, for every detection rule dr_d such that $dr_d.producesState = state$, the physical properties required by the input parameters par_p of dr_d, minus the ones already provided by so_n, are returned.

This way, so_n can either be replaced, or altered by adding new sensors, such that all par_p of at least one dr_d can be provided. The applicability of this strategy depends on the number of smart objects to alter or replace, and on their accessibility, cost, and ownership. For example, suppose that the truck $truckVA789TY$ is only equipped with an indoor humidity sensor. In this case, no correlation exists between the truck being on the move and the indoor humidity. Therefore, no detection rule dr_d uses humidity data to infer if the truck is *moving*. So, to infer that state, either additional sensors have to be installed, or the truck has to be replaced.

6   Related Work

To detect when activities are executed based on data coming from sensors, Baumgrass et al. [3] integrate a BPMN engine with a Complex Event Processor (CEP). The BPMN language is extended with Process Event Monitoring Points (PEMPs), events that, when received, determine the activation or termination of activities. The CEP is then responsible for deriving PEMPs from sensor data. An architecture that implements this solution is proposed by Bülow et al. [5]. With respect to our work, the concept of artifact is absent. Therefore, the BPMN process model alone is not sufficient to understand which physical objects influence the execution of the process. Gnipieba et al. [9], on the other hand, propose a collaboration hub, driven by a Guard-Stage-Milestone (GSM) [13] artifact-centric process model, to monitor the execution of multi-party processes. The GSM process model directly contains rules to determine, based on sensor values coming from smart objects, the activation and termination of activities. Our work differentiates from [9] by decoupling from the process model the rules required to infer the state of the artifacts. This way, the process model uses only information on the state of the artifacts to determine when activities are run. As such, the process model does not have to be changed when multiple executions of the process, each one differing from the others in terms of smart objects and their capabilities, take place.

Concerning the usage of artifacts to monitor a process, Meyer et al. [16] derive SQL queries from a process model annotated with artifacts. Such queries are then used to manipulate the artifacts, which are represented in a database, and to detect when activities can be executed. While this work has some similarities to our approach, we offer a complementary vision: [16] mainly focuses on virtual artifacts (i.e., invoices, purchase orders, etc.), whereas we concentrate on physical artifacts, without requiring their attributes or states, which constantly change, to be stored in a database. Other research work, such as [7] or [8], focuses on the cause-effect relationships between artifacts and activities. However, their goal is to rely on the execution of activities to infer how and when the artifacts change, while we do the exact opposite.

Besides FIESTA-IoT [1] and the Physics Domain Ontology [12], ontologies supporting the IoT have also been proposed by Nambi et al. [19], Xu et al. [23], and Wang et al. [21], just to name a few. However, none of them is fully suited for describing the characteristics of an artifact-driven monitoring infrastructure. In particular, only FIESTA-IoT explicitly supports roaming smart objects, i.e., smart objects that have no fixed location. [19,21] make a one-to-one association between sensors and smart objects. By doing so, they keep the sensing infrastructure disjoint from the artifacts they monitor, thus not fully embracing the IoT paradigm [17], where physical objects *become* smart objects.

7   Conclusion and Future Work

In this paper we have proposed how to enable an artifact-driven monitoring architecture by exploiting two ontologies to describe the capabilities of a monitoring infrastructure. These ontologies extend existing ontologies related to the IoT domain by adding information relevant for the monitoring purposes. As discussed in the paper, the ontologies can be used to measure the process model monitorability with respect to the available monitoring infrastructure. Moreover, we also investigated how the ontologies can also be used to improve the monitorability of a process by suggesting to the process designer which modifications could be done to the process model or to the monitoring infrastructure.

As the ontologies are now manually queried, future work will concentrate on building an application on top of them able to assist the process designer during the process improvement phase. Moreover, we want to make possible to automatically detect the minimum set of modifications in the process model and/or infrastructure that maximizes the monitorability of a process.

Acknowledgments. This work has been partially funded by the Italian Project ITS Italy 2020 under the Technological National Clusters program.

References

1. Agarwal, R., Fernandez, D.G., Elsaleh, T., Gyrard, A., Lanza, J., Sánchez, L., Georgantas, N., Issarny, V.: Unified IoT ontology to enable interoperability and federation of testbeds. In: WF-IoT 2016, pp. 70–75. IEEE Computer Society (2016)
2. Baresi, L., Meroni, G., Plebani, P.: On handling business process anomalies through artifact-based modeling. In: CAiSE 2016 Forum, pp. 9–16. CEUR-WS.org (2016)
3. Baumgrass, A., Herzberg, N., Meyer, A., Weske, M.: BPMN extension for business process monitoring. In: EMISA 2014, pp. 85–98. GI (2014)
4. vanden Broucke, S.K.L.M., Caron, F., Lismont, J., Vanthienen, J., Baesens, B.: On the gap between reality and registration: a business event analysis classification framework. Inf. Technol. Manag. **17**(4), 393–410 (2016)
5. Bülow, S., et al.: Monitoring of business processes with complex event processing. In: Lohmann, N., Song, M., Wohed, P. (eds.) BPM 2013. LNBIP, vol. 171, pp. 277–290. Springer, Cham (2014). doi:10.1007/978-3-319-06257-0_22
6. Compton, M., et al.: The SSN ontology of the W3C semantic sensor network incubator group. J. Web Semant. **17**, 25–32 (2012)
7. Eid-Sabbagh, R.-H., Hewelt, M., Meyer, A., Weske, M.: Deriving business process data architecturesfrom process model collections. In: Basu, S., Pautasso, C., Zhang, L., Fu, X. (eds.) ICSOC 2013. LNCS, vol. 8274, pp. 533–540. Springer, Heidelberg (2013). doi:10.1007/978-3-642-45005-1_43
8. Eshuis, R., Gorp, P.: Synthesizing object life cycles from business process models. In: Atzeni, P., Cheung, D., Ram, S. (eds.) ER 2012. LNCS, vol. 7532, pp. 307–320. Springer, Heidelberg (2012). doi:10.1007/978-3-642-34002-4_24
9. Gnimpieba, Z.D.R., Nait-Sidi-Moh, A., Durand, D., Fortin, J.: Using Internet of Things technologies for a collaborative supply chain: application to tracking of pallets and containers. Procedia Comput. Sci. **56**, 550–557 (2015)
10. Guceglioglu, A.S., Demirors, O.: A process based model for measuring process quality attributes. In: Richardson, I., Abrahamsson, P., Messnarz, R. (eds.) EuroSPI 2005. LNCS, vol. 3792, pp. 118–129. Springer, Heidelberg (2005). doi:10.1007/11586012_12
11. Gyrard, A., Datta, S.K., Bonnet, C., Boudaoud, K.: Cross-domain Internet of Things application development: M3 framework and evaluation. In: FiCloud 2015, pp. 9–16. IEEE Computer Society (2015)
12. Hachem, S., Teixeira, T., Issarny, V.: Ontologies for the Internet of Things. In: MDS 2011, pp. 3:1–3:6. ACM (2011)
13. Hull, R., et al.: Introducing the guard-stage-milestone approach for specifying business entity lifecycles. In: Bravetti, M., Bultan, T. (eds.) WS-FM 2010. LNCS, vol. 6551, pp. 1–24. Springer, Heidelberg (2011). doi:10.1007/978-3-642-19589-1_1
14. Ly, L.T., Maggi, F.M., Montali, M., Rinderle-Ma, S., van der Aalst, W.M.: Compliance monitoring in business processes: functionalities, application, and tool-support. Inf. Syst. **54**, 209–234 (2015)
15. Meroni, G., Di Ciccio, C., Mendling, J.: Artifact-driven process monitoring: dynamically binding real-world objects to running processes. In: Proceedings of the Forum and Doctoral Consortium Papers Presented at the 29th International Conference on Advanced Information Systems Engineering, CAiSE 2017, Essen, Germany, June 12-16, pp. 105–112 (2017). http://dblp2.uni-trier.de/rec/bibtex/conf/caise/MeroniCM17

16. Meyer, A., Pufahl, L., Fahland, D., Weske, M.: Modeling and enacting complex data dependencies in business processes. In: Daniel, F., Wang, J., Weber, B. (eds.) BPM 2013. LNCS, vol. 8094, pp. 171–186. Springer, Heidelberg (2013). doi:10. 1007/978-3-642-40176-3_14
17. Miorandi, D., Sicari, S., Pellegrini, F.D., Chlamtac, I.: Internet of Things: vision, applications and research challenges. Ad Hoc Netw. **10**(7), 1497–1516 (2012)
18. Musen, M.A.: The protégé project: a look back and a look forward. AI Matters **1**(4), 4–12 (2015)
19. Nambi, S., Sarkar, C., Prasad, R.V., Biswas, A.R.: A unified semantic knowledge base for IoT. In: WF-IoT 2014, pp. 575–580. IEEE Computer Society (2014)
20. Uschold, M., Gruninger, M.: Ontologies: principles, methods and applications. Knowl. Eng. Rev. **11**(2), 93–136 (1996)
21. Wang, W., De, S., Tönjes, R., Reetz, E.S., Moessner, K.: A comprehensive ontology for knowledge representation in the Internet of Things. In: TrustCom 2012, pp. 1793–1798. IEEE Computer Society (2012)
22. Weske, M.: Business Process Management - Concepts, Languages, Architectures, 2nd edn. Springer, Heidelberg (2012). doi:10.1007/978-3-642-28616-2
23. Xu, Y., Zhang, C., Ji, Y.: An upper-ontology-based approach for automatic construction of IOT ontology. IJDSN **10**, 1–17 (2014)

A Quantitative Study of the Link Between Business Process Management and Digital Innovation

Amy Van Looy$^{(\boxtimes)}$ (iD)

Department of Business Informatics and Operations Management,
Faculty of Economics and Business Administration, Ghent University,
Tweekerkenstraat 2, 9000 Ghent, Belgium
Amy.VanLooy@UGent.be

Abstract. The current digital era is characterized by increasing globalization and a fast evolution in new technologies (e.g. social media, mobile, cloud, big data analytics, Internet of Things and smart devices). Since organizations are exponentially challenged to achieve business results more effectively and efficiently, topics such as process performance and the constant search for optimizations and transformations are key. Business Process Management (BPM) and digital innovation are both ways for organizations to constantly improve, change and excel. Nonetheless, more research is needed on the intersection between these two approaches. While current literature acknowledges a link between BPM and digital innovation, this study digs deeper into the digital innovation strategies that organizations apply to incorporate technological transformations into their business processes. Based on our survey findings, we open the discussion about the strength of the assumed relationship between BPM and digital innovation, and which implications can be drawn.

Keywords: Process change management · Lifecycle management · Capability · Critical success factor · Adoption · Digital innovation · Digital transformation

1 Introduction

Managing and improving business processes (in short: Business Process Management or BPM) is valuable for many organizations [1]. BPM intends to align business processes to corporate objectives and strategies for obtaining more efficient and effective business processes and achieving better business results. Throughout the years, BPM has evolved into a holistic discipline that focuses on the entire portfolio of business processes in and between organizations, in which digital innovation plays an increasing role [2]. Modern businesses thus require agility, flexibility and innovation instead of only continuous improvements, automation and standardization [3].

Since the years 2000s, the role of computers and the Internet has tremendously increased. Because hardware is becoming cheaper and more powerful, organizations profit more from managers being able to think of new or improved business processes [4]. The digitalization of today's work environment implies the adoption of digital innovation

© Springer International Publishing AG 2017
J. Carmona et al. (Eds.): BPM Forum 2017, LNBIP 297, pp. 177–192, 2017.
DOI: 10.1007/978-3-319-65015-9_11

based on new technologies, such as social media, mobile and cloud solutions, big data analytics, RFID, sensors, Internet of Things and smart devices. Given the fast pace in which new technologies change, digital innovation exerts a continuous influence on organizations and their business processes [5–7]. This influence on the digitalizing economy has multiple implications. For instance, the fast emergence of new technologies has increased the speed of business innovations and transformations [7, 8]. Although a proper use of new technologies is not evident, it can also offer new knowledge and stronger insights into an organization's way of working [6]. Nonetheless, the more technologies become user-friendly and competitive, the more important it is to incorporate them in an organization's strategies and business processes [7]. These influences also demonstrate a more essential role of CIOs in organizations.

Consequently, similar to the reengineering wave in the 1990s, a shift can now be observed within the BPM discipline from an automation logic to an innovation logic [9]. As such, organizations are preparing themselves to survive and/or to grow in current or other markets [8]. In addition to new technologies, customers should take an increasingly prominent role. For instance, social media help customers give feedback very quickly and organizations adopt social Customer Relationship Management systems. Hence, more than ever before, business processes should be properly aligned to customer requirements [2, 5]. In response, the BPM discipline starts accepting the notions of Customer Process Management and of organizational ambidexterity, which combines exploitative and explorative BPM [10]. While the current body of knowledge acknowledges a link between BPM and digital innovation [8], more research is needed on concrete theories, models or applications [10].

To partly fill this gap, our research questions (RQs) are as follows.

- **RQ1.** Does a relationship exist between BPM and digital innovation, and if so, what is its strength?
- **RQ2.** If a BPM-digital innovation relationship exists (RQ1), to which extent do digital innovation strategies differ between a higher and lower BPM adoption?
- **RQ3.** If a BPM-digital innovation relationship exists (RQ1), which BPM-specific capabilities or critical success factors contribute more to this relationship?

This study extends the body of knowledge with 1/ quantitative evidence from a survey that takes a 2/ more refined view on BPM and digital innovation based on existing frameworks and categorizations. We explore those factors that contribute to the assumed relationship to gain more insight and open the discussion about the role of BPM in the digitalizing economy. The practical relevance of our study is providing organizations with a general idea of the current situation. On the longer run, findings from this study combined with similar studies in the future will help formulate guidelines on the implementation of digital innovation to enhance BPM adoption, as well as developing theories, models and applications that fit different business contexts.

The remainder is structured as follows. We provide the theoretical background of BPM and digital innovation in Sect. 2. Section 3 describes our research method. Afterwards, the results are presented (Sect. 4) and discussed (Sect. 5).

2 Theoretical Background

2.1 Business Process Management

BPM is defined as *"the art and science of overseeing how work is performed in an organization to ensure consistent outcomes and to take advantage of improvement opportunities"* [11: p. 1]. It is assumed that each business process advances through iterations in a lifecycle, also known as the Plan-Do-Check-Act cycle. Although such iterative lifecycle symbolizes the idea of continuous improvement, process optimizations may range from smaller, incremental changes to larger, more radical improvements. While continuous process improvements have been advocated by (total) quality management thinkers for decades, the idea of radical process changes only grew since the 1990s due to IT opportunities and globalization with increased competition. The main representatives of radical changes were [12] with business process reengineering (BPR) and [13] with process innovation. In the 2000s, similar triggers (i.e. IT and globalization) grew stronger with new technologies and e-business/e-commerce.

Previous studies have shown that BPM can lead to better business (process) performance and competitiveness [14–17]. The critical success factors to advance in BPM have been quantitatively measured in multiple maturity models, which aim to gradually support organizations by providing step-by-step guidance [15, 16, 18]. Not only has their high number made it difficult to choose one maturity model, also their difference in scope did [19]. In other words, different maturity types exist [20]. For instance, maturity models may focus on a different set of business processes (i.e. individual processes versus the entire process portfolio in an organization) as well as a different set of critical success factors (i.e. capability areas limited to the lifecycle versus also addressing organizational areas such as culture and structure). The latter differentiates the narrow view on BPM from the more holistic view, which is also called Business Process Orientation (BPO) [15, 20]. Hence, in an attempt to summarize all BPM-related critical success factors, [20] propose a conceptual framework that is based on a literature review, grounded in the process lifecycle and organizational management theories, and is empirically validated by 69 maturity models.

A misconception exists that all organizations should blindly strive for the highest BPM maturity levels. Instead, each organization should decide on its optimal level considering its business context. A business context is generally examined in contingent or context-aware studies by empirically relating BPM maturity to factors such as sector or size [21, 22]. Other approaches are making the BPM-related critical success factors more case-specific [23] or prioritize improvement alternatives based on milestones [24]. Such studies explain, for instance, why most public sector organizations have a lower BPM maturity than organizations in market-competitive sectors.

The reasoning above is, however, not applicable for explaining the link between BPM and digital innovation, since many public sector organizations are also using new technologies (e.g. to become smart cities). Although one may think that higher innovativeness relates to lower BPM maturity due to BPM's emphasis on continuous improvements, automation and standardization, [14] show empirical evidence for the contrary. Additionally, [25] argue that organizations with a higher BPM maturity level can take a more standards-based approach as well as a custom-made approach. Hence, additional explanations are needed about the way BPM is applied, and about a (possibly new) role of BPM due to the increasing importance of digital innovation.

2.2 Digital Innovation

Digital innovation is defined as "*a product, process, or business model that is perceived as new, requires some significant changes on the part of adopters, and is embodied in or enabled by IT*" [7: p. 330]. Its origins can be internal or external (i.e. open innovation with stakeholders) [27]. Regarding BPM, digital innovation implies creating novel business processes or substantially changing existing processes by IT [7]. [28] further specifies two categories of digital innovation in BPM: (1) digital changes in the way of working, and (2) creations of more intelligent processes. Concerning the origins, [27] suggest that BPM profits more from external than internal (digital) innovations because valuable relationships are better enablers than money.

Although digital innovation is seen as increasingly important, it is not easy to realize, control and predict [26]. The decision to invest in digital innovation can depend on strategic aspects, e.g. earlier business decisions, the expected opportunities in the future, and the expected speed and impact of new IT [29]. Other factors affecting the choice for a digital strategy are related to the business context, e.g. industry growth, concentration and turbulence [30]. [31] agree by emphasizing that a successful digital innovation is rather driven by strategy, culture and talent development than merely using new technologies. Additionally, [30] propose two digital strategy types: general IT investments and IT outsourcing investments. Per strategy, it is important to observe whether the budget has substantially increased, and this 1/throughout the years and 2/relative to competitors. Given strategy's crucial role, this work will take digital business strategies as a first proxy for digital innovation.

Similar to BPM maturity models, organizations may feel the need for support by diagnostic tools or frameworks to help realize a digital innovation. For instance, the overall stages through which any digital innovation evolves appear to be similar: discovery, development, diffusion, and impact [7]. A digital maturity model is presented by [32], albeit for telecommunications providers. [26] propose a generic diagnostic tool with five items across three groups: (1) a product group with user experience and value proposition, (2) an environment group with digital evolution scanning, (3) an organization group with skills- and improvisation-related items. Since an established measure for digital innovation is missing, this generic index will be a second proxy.

3 Methodology

We collected data in November 2016. The survey contained closed questions to decrease a researcher bias. For reasons of data quality and to ensure a higher response rate, we opted for an oral questionnaire. This means that the respondents filled out the survey questions together with a Master student in IT management who followed a mandatory BPM course. The oral interviews were limited to reading and (if necessary) explaining the questions. This approach resulted in a relatively high response rate of 38.14%. The average survey duration was 30 min.

To further ensure data quality, respondents could only participate if they satisfied five conditions: (1) fulfill a managerial function, (2) have a total seniority of at least five years, (3) have a seniority of at least two years in the current organization, (4) have an interest in BPM, and (5) understand English fluently. The respondents were mainly selected by e-mail to request for a face-to-face interview (e.g. by randomly screening LinkedIn profiles) or respondents lived in the neighborhood of the Master students. In total, 148 surveys were registered, of which 133 had reached the end of the survey and 131 had answers to all survey questions. The final sample thus covered 131 (mainly West European) managers, each belonging to a different organization (Appendix A). Regarding external validity, we acknowledge the limitation of generalizing our findings to all organizations worldwide. Nonetheless, the results help explore and open the discussion for further investigation of the topic.

The main variables were related to a BPM index [20], digital strategies [30], and a digital innovation index [26]. To increase internal validity, the operationalization was primarily based on recognized measurement instruments or frameworks (Table 1).

Table 1. An overview of the surveyed variables.

Variable	Literature	Number of items	Operationalization	Measurement level
BPM index	[20]	65	7-point-Likert scales (1 = strongly agree; 7 = strongly disagree)	• Ordinal per item • Interval per capability (= average of items) • Interval as index (= sum of averages)
BPM control variable	Self-administered	1	Score out of 10 (1 = not process-oriented; 10 = fully process-oriented)	Interval per item
New IT types	Sect. 1	8	Binary (yes/no)	Nominal per item
Digital strategy types	[30]	4	7-point-Likert scales (1 = strongly agree; 7 = strongly disagree)	Ordinal per item
Digital innovation index	[26]	15	7-point-Likert scales (1 = strongly agree; 7 = strongly disagree)	• Ordinal per item • Interval per capability (= average of items) • Interval as index (= sum of averages)
Digital innovation control variable	Self-administered	1	Score out of 10 (1 = no digital innovation; 10 = fully digitally innovated)	Interval per item

For those variables that combined items, we calculated a Cronbach's alpha of above 0.7, indicating data reliability (i.e. $\alpha = 0.940$ for the capabilities in the BPM index; $\alpha = 0.884$ for the capabilities in the digital innovation index). We also introduced control variables. Correlation coefficients indicated a moderate to strong linear relationship for the BPM index and its control variable (Pearson's $r = -0.626$; Spearman's rho $= -0.552$; Kendall's tau_b $= -0.426$; $P = 0.000$) as well as for the digital innovation index and its control variable (Pearson's $r = -0.671$; Spearman's rho $= -0.613$; Kendall's tau_b $= -0.486$; $P = 0.000$). These correlation coefficients are negative, since the scales for the indices and the control variables were opposite (Table 1). As such, further evidence is observed for internal consistency and data reliability.

4   Results

4.1   Results for RQ1

Almost all (i.e. except three) organizations were using at least one new technology. The top-5 was social media (76.3%), mobile (75.6%), cloud solutions (67.2%), big data/business intelligence (56.5%), and Internet of Things/smart devices (42.7%).

Due to the relatively small sample size (N = 131), the tests of Kolmogorov-Smirnov and Shapiro-Wilk were less powerful and normality was not assumed for all variables. We thus used non-parametric correlation tests: Spearman's rho and Kendall's tau_b.

Table 2 shows that both correlation tests presented statistically significant, positive relationships between the BPM index and the other variables. The correlation with the degree of digital innovation was moderate ($P < 0.001$), with IT investment strategies of increasing budget and relative budget moderate to low ($P < 0.001$), and with IT outsourcing strategies of increasing budget and relative budget rather low to weak ($P < 0.050$ and $P < 0.001$). Since the literature shows no decisive evidence for causal relationships, we did not perform a regression analysis.

Table 2. The correlation tests with respect to the BPM index (N = 131).

Correlation	Spearman's rho	Kendall's tau_b
BPM index * Digital innovation index	0.516***	0.360***
BPM index * IT investments (increasing budget)	0.371***	0.278***
BPM index * IT investments (relative budget)	0.294***	0.220***
BPM index * IT outsourcing (increasing budget)	0.255**	0.186**
BPM index * IT outsourcing (relative budget)	0.291***	0.215***

(NS $P > 0.100$; * $P < 0.100$; ** $P < 0.050$; *** $P < 0.001$)

4.2   Results for RQ2

The hypotheses for RQ2 were:

- H_0: The **degree** and **strategies** for **digital innovation** are **independent** of the degree of **BPO adoption**.

- H_a: The **degree** and **strategies** for **digital innovation** significantly **differ** among the degree of **BPO adoption**.

We recoded the continuous BPM index (i.e. with scores ranging from 13 to 91) into a categorical variable with three groups (i.e. 13 to 38.9 for higher BPM adoption; 39 to 64.9 for medium BPM adoption; 65 to 91 for lower BPM adoption). The continuous digital innovation index was also recoded (i.e. with scores from 5 to 14.9 for a higher degree of digital innovation; 15 to 24.9 for a medium degree; 25 to 35 for a lower degree) to allow a first screening of the hypotheses based on a cross tabulation. Table 3 concerns a 3 × 3 table showing unequal sample sizes among the groups and only a few respondents in the lower categories (i.e. four with a lower BPM adoption and six with a lower degree of digital innovation). Hence, instead of the Chi-square test, we calculated the Fisher-Freeman-Halton Exact test with a value of 44.898 and four degrees of freedom. This indicated that the recoded BPM index and the recoded digital innovation index appear to be dependent ($P = 0.000$).

Table 3. The cross tabulation of digital innovation by BPM (based on recoded indices).

		BPM index (recoded)			Total
		Higher BPM adoption	Medium BPM adoption	Lower BPM adoption	
Digital innovation index (recoded)	Higher degree of digital innovation	**49**	5	0	54
	Medium degree of digital innovation	33	**37**	1	71
	Lower degree of digital innovation	1	2	**3**	6
Total		83	44	4	131

Table 4 summarizes the descriptive statistics per recoded BPM adoption group.

To verify homogeneity of variance, we calculated the Levene's test in its traditional version and non-parametric version, given the fact that our data were not normally distributed [33]. While the non-parametric version of the Levene's test gave evidence for unequal variances among all main variables ($P < 0.050$), the traditional Levene's test suggested a combination of equal variances for the digital innovation index, the IT investment strategies and the increasing budget for IT outsourcing ($P > 0.050$ each), as well as unequal variances for the recoded digital innovation index ($P < 0.050$) and the relative budget for IT outsourcing ($P < 0.100$). Based on the verified assumptions for normality, sample sizes and homogeneity of variance, we opted for both the Kruskal-Willis H rank test and the ANOVA Welch's F test. The non-parametric Kruskal-Wallis test can deal with non-normality but requires homogeneity of variance, while the parametric Welch's test does not require equal variances or equal group sizes and is relatively robust for non-normality [34]. Previous research showed that parametric tests could still be powerful for non-normal data with small sample sizes [34, 35].

Table 4. Descriptive statistics for the main variables, per BPM adoption group (N = 131).

Variables	BPM index (recoded)					
	Higher BPM adoption		Medium BPM adoption		Lower BPM adoption	
	Mean	Median	Mean	Median	Mean	Median
Digital innovation index	14.245	14.000	18.205	18.667	27.417	27.167
Digital innovation index (recoded)	1.422	1.000	1.932	2.000	2.750	3.000
IT investments (increasing budget)	2.170	2.000	3.050	3.000	4.000	3.500
IT investments (relative budget)	3.160	4.000	3.910	4.000	4.500	4.000
IT outsourcing (increasing budget)	2.830	2.000	3.340	3.000	4.250	4.000
IT outsourcing (relative budget)	3.580	4.000	4.140	4.000	4.500	4.000

4.2.1 ANOVA Welch's F Test

The ANOVA Welch's F test showed that at least one BPM group differs from another for the digital innovation index (F = 30.337; df1 = 2; df2 = 5.823; $P < 0.001$), its recoded index (F = 26.023; df1 = 2; df2 = 8.194; $P < 0.001$), the increased IT investments over time (F = 5.750; df1 = 2; df2 = 7.882; $P < 0.050$) and relative to competitors (F = 4.529; df1 = 2; df2 = 8.075; $P < 0.050$), but not for IT outsourcing strategies ($P > 0.100$). To identify the BPM groups among which a difference is expected, we performed the Games-Howell post-hoc test (Table 5) under the assumption of unequal variances and unequal sample sizes. This test is robust for non-normality [36].

From Table 5, it can be expected that organizations with a higher BPM adoption also have a higher degree of digital innovation than organizations with a medium or lower BPM adoption ($P < 0.001$ and $P < 0.050$). This finding is for both the digital innovation index and its recoded index. For the recoded digital innovation index, it was found that organizations with a medium BPM adoption have a higher degree of digital innovation than those with a lower BPM adoption ($P < 0.100$). The data also suggested that organizations with a higher BPM adoption are investing more in IT over the years and more than the competitors compared to the medium BPM adoption ($P < 0.050$).

The estimated increases in units, shown in Table 5, are based on pairwise comparisons of the means. The negative values were due to the inverse scales (see Table 1).

Table 5. Post-hoc testing with mean differences related to the ANOVA Welch's test results.

Games-Howell post-hoc testing for multiple (pairwise) comparisons	BPM index (recoded)		
	Higher vs Medium BPM adoption	Higher vs Lower BPM adoption	Medium vs Lower BPM adoption
Digital innovation index	−3.960***	−13.17**	−9.212[NS]
Digital innovation index (recoded)	−0.510***	−1.328**	−0.818*
IT investments (increasing budget)	−0.877**	−1.831[NS]	−0.955[NS]
IT investments (relative budget)	−0.752**	−1.343[NS]	−0.591[NS]

(NS $P > 0.100$; * $P < 0.100$; ** $P < 0.050$; *** $P < 0.001$)

For instance, for the digital innovation index, the mean score for organizations with a higher degree of BPM adoption is expected to be four units (of digital innovation) higher than for organizations with a medium degree of BPM adoption, and 13 units higher than for organizations with a lower degree of BPM adoption. For the IT investment strategies, the estimated differences are circa one unit on a 7-point Likert scale between a higher and medium BPM adoption. This corresponds to Table 4.

4.2.2   Kruskal-Wallis H Rank Test

The Kruskal-Wallis test showed that at least one BPM adoption group differs from another for the digital innovation index ($\chi^2 = 38.952$; df = 2; $P < 0.001$), its recoded index ($\chi^2 = 35.401$; df = 2; $P < 0.001$), the IT investments strategies with increasing budget ($\chi^2 = 14.436$; df = 2; $P < 0.001$) and relative budget ($\chi^2 = 7.920$; df = 2; $P < 0.050$), and the IT outsourcing strategies with increasing budget ($\chi^2 = 5.042$; df = 2; $P < 0.100$) and relative budget ($\chi^2 = 4.636$; df = 2; $P < 0.100$). The degree of digital innovation and digital strategies are thus expected to be different between the BPM adoption groups. This finding is in line with the Welch's tests, although the IT outsourcing strategies were now statistically significant. To know which groups differ, we recalculated the Kruskal-Wallis rank test based on one degree of freedom (Table 6). Since the recoded BPM index had three groups (i.e. ≤ 4 groups), we did not use a Bonferroni correction.

From Table 6 follows that organizations with a higher BPM adoption have a higher degree of digital innovation than organizations with a medium or lower BPM adoption ($P < 0.001$). Similarly, organizations with a medium BPM adoption have a higher degree of digital innovation than those with a lower BPM adoption ($P < 0.050$). The same findings were observed for the recoded digital innovation index ($P < 0.001$). Regarding the IT investment strategies, organizations with a higher BPM adoption are

Table 6. Post-hoc testing with mean ranks related to the Kruskal-Wallis test results.

Non-parametric one-way ANOVA for multiple (pairwise) comparisons	BPM index (recoded)					
	Higher vs Medium BPM adoption		Higher vs Lower BPM adoption		Medium vs Lower BPM adoption	
	χ^2	Effect	χ^2	Effect	χ^2	Effect
Digital innovation index	30.200***	23.97%	10.797***	12.56%	8.373**	17.82%
Digital innovation index (recoded)	26.605***	21.12%	11.653***	13.55%	11.536***	24.55%
IT investments (increasing budget)	11.587***	9.20%	4.363**	5.07%	1.077[NS]	–
IT investments (relative budget)	6.835**	5.43%	1.796[NS]	–	0.237[NS]	–
IT outsourcing (increasing budget)	3.278*	2.60%	2.361[NS]	–	0.804[NS]	–
IT outsourcing (relative budget)	4.264**	3.38%	0.731[NS]	–	0.006[NS]	–

(NS $P > 0.100$; * $P < 0.100$; ** $P < 0.050$; *** $P < 0.001$)

investing more in IT over the years and more than their competitors compared to the medium BPM adoption group ($P < 0.001$ and $P < 0.050$). Organizations with a higher BPM adoption are also investing more in IT over the years compared to those with a lower BPM adoption ($P < 0.050$). We also found that organizations with a higher BPM adoption are spending more on IT outsourcing over the years and more than their competitors compared to the medium BPM adoption group ($P < 0.100$ and $P < 0.050$).

Since the pairwise comparisons were based on mean ranks instead of means, the estimated increase in units is less evident to interpret. In addition, a generalization to differences in medians is less appropriate since the distributions had different shapes and variabilities. Thus, we looked at the effect size estimates (eta-squared), i.e. which percentage of the variability in ranked scores is accounted for by the BPM adoption groups. Table 6 illustrates that the effect size estimates were relatively high for the digital innovation index and its recoded index (i.e. between 12.56% and 24.55%), and decent for the IT investment strategies (i.e. between 5% and 10%). The more the estimates are closer to or less than five, the less effect is expected. This was already true for the increasing IT budget between a higher and lower BPM adoption (5.07%), but especially for the IT outsourcing strategies with relative budget (3.38%) and increasing budget (2.60%). IT outsourcing was also not significant in Sect. 4.2.1.

4.3   Results for RQ3

Table 7 shows that positive correlations exist between the digital innovation index and all capabilities in the BPM index ($P < 0.001$). The BPM capabilities with the highest correlations (rho > 0.4; tau_b > 0.3) were: **(1) process-oriented governance bodies, (2) Act-phase of the process lifecycle, (3) top management commitment, (4) process-oriented skills and training.** While these four correlations were moderate, the other correlations had weaker relationships but still moderate to low (rho > 0.3; tau_b > 0.2). The lowest value was for process-oriented appraisals and rewards (HR).

Table 7. The correlation tests for the BPM capabilities in the BPM index (N = 131).

Correlation of the digital innovation index with:	Spearman's rho	Kendall's tau_b
Plan-phase of the process lifecycle	0.365***	0.257***
Do-phase of the process lifecycle	0.331***	0.240***
Check-phase of the process lifecycle	0.358***	0.261***
Act-phase of the process lifecycle	**0.462***	**0.337***
Strategic alignment	0.344***	0.251***
External relationships	0.331***	0.240***
Process-oriented roles and responsibilities	0.392***	0.274***
Process-oriented skills and training	**0.503***	**0.362***
Process-oriented values, attitudes and behaviors	0.352***	0.242***
Process-oriented appraisals and rewards	0.314***	0.220***
Top management commitment	**0.488***	**0.357***
Process-oriented organization chart	0.336***	0.246***
Process-oriented governance bodies	**0.438***	**0.317***

(NS $P > 0.100$; * $P < 0.100$; ** $P < 0.050$; *** $P < 0.001$)

5   Discussion

5.1   Discussion for RQ1

While BPM scholars recognize the importance of digital innovation [8], the current body of knowledge is extended with more quantitative evidence of the strength of this relationship. Given the relatively small sample size (N = 131), our data rather explore how a more refined view can be taken by means of an overall digital innovation index [26] and digital innovation strategies [30]. While the observed relationship between BPM and **digital innovation in general** turned out to be the strongest, it is still rather moderate. Similar to previous research [8, 14], we can state that BPM and digital innovation may support one another. Particularly, digital innovation can support business processes (and thus BPM) with new IT being a driver and an implementer for process changes. Likewise, BPM can stimulate digital innovation, and IT-enabled business processes can make organizations more innovative and competitive. Also [28] pointed out that many emerging technologies are implemented in BPM. The moderate degree of this relationship can be explained by the fact that overprocessing and rigid or too strict business processes may also kill creativity and innovativess, or when BPM supports efficiency but efficiency running in the wrong direction.

BPM appears to be linked to **IT investments** rather than **IT outsourcing**. The latter is in line with the BPR idea of drastic IT-enabled process changes and process innovations from the 1990s [12, 13]. Furthermore, the costs for implementing (new) IT (such as social media) are decreasing given their omnipresence in today's society, whereas the role of the IT department is increasing. Another explanation might be that organizations particularly apply BPM to their core processes, which they want to control themselves, while process outsourcing is rather applicable to non-core (e.g. supporting) processes or core processes for which partners have more expertise.

5.2   Discussion for RQ2

Careful conditions were set to participate in our sample (Sect. 3), such as having an interest in BPM. This may explain why our data primarily reflect organizations having a higher or medium degree of BPM adoption. Still, the analyses of variance accepted the hypotheses that the degree and strategies for digital innovation significantly differ among the BPM adoption groups. Again, the strongest relationship exists between BPM adoption and digital innovation, and this across the higher, medium and lower BPM adoption groups. The differences in IT strategies among the BPM adoption groups are also more valid for IT investments than IT outsourcing.

First, the findings quantitatively agree with [9] that organizations can use digital opportunities as the basis of BPM, i.e. to innovate, reengineer or redesign business processes, and thus achieving a higher degree of BPM adoption. Similarly, organizations active in BPM possibly see more opportunities in digital innovation. **Hence, organizations with a higher BPM adoption might invest faster in digital innovation compared to organizations with a medium and lower adoption. Likewise, organizations with a medium BPM adoption appear to invest more in digital**

innovation than organizations with a lower BPM adoption. One explanation is that process modelling can reveal pain points in business processes, which may be solved using digital innovation. This, however, does not imply that organizations with a lower BPM adoption cannot invest in digital innovation. Another explanation is that organizations with a higher BPM adoption might rely more on a long-term vision compared to organizations with a lower BPM maturity. BPM maturity models are typically used to identify improvement areas and to apply BPM in a more focused way. Given this positive relationship, BPM maturity models may indirectly encourage more advanced digital innovation strategies as an additional trigger for using those models.

Also regarding the IT investment strategies, the results give quantitative evidence that BPM is shifting from an automation logic to an innovation logic [9]. **In other words, organizations with a higher BPM adoption appear to invest more in their IT infrastructure compared to organizations with a lower BPM adoption, and especially compared to organizations with a medium BPM adoption.** One explanation is that organizations with a higher BPM adoption rely more on their IT department to introduce and advance in BPM. By implementing new technologies, organizations might reduce costs, gain a competitive advantage (or at least not lagging too far behind) or better satisfy customer requirements. Another explanation is that organizations with a higher BPM adoption might rely more on social media for improving their work, both internally and with customers and stakeholders, or work more in a paperless office. Therefore, it is recommended to regularly update the organization's IT infrastructure. Previous studies on critical success factors for BPM also consider the importance of IT investments [23] and IT alignment [37]. [23] suggests a task-technology fit and alignment or contingency between the level of IT investments, corporate strategies and investment level in business processes and BPM. Also [37] refers to process alignment as a fit between an organization's processes and its institutional elements, e.g. IT alignment, given that IT can enable (core) processes.

Finally, we note that **the findings for IT outsourcing strategies are less significant, and a possible difference is only expected to some extent between a higher and medium BPM adoption.** We refer to Sect. 5.1 for a possible explanation.

5.3   Discussion for RQ3

RQ3 focused on individual BPM capabilities or critical success factors, because organizations with a higher BPM adoption have more developed capabilities [20]. While positive relationships were observed for all capabilities, four areas appeared to contribute more to digital innovation. The first refers to **process-oriented governance bodies**, with a program manager who coordinates all BPM projects and a Center of Excellence (CoE) that shares expertise on BPM methods and techniques. The CoE is thus related to support and training. Organizations can also coordinate mechanisms for improvisation and flexibility of employees to increase (process) performance.

The second capability is related to the **Act-phase of the process lifecycle**, which ultimately focuses on (IT-enabled) process change and innovation. By following each lifecycle phase, managers can observe, listen and involve employees, leading to novel ideas from bottom-up. Ideas of co-workers are potentially more easily accepted and implemented. Regular feedback from employees and customers may affect employee satisfaction, which in turn is translated into customer satisfaction.

The third capability is **top management commitment**, and emphasizes the importance of a Chief Process Officer and top managers stimulating a process-oriented way of working (instead of traditional, vertical departments and functions). Moreover, it is crucial that top managers transfer these ideas to their employees. Top managers also have a better overview of their business and corporate vision, mission and strategies. Since long-term decisions prevail more on top management level, top managers should walk the talk. Hence, the willingness of top management and employees are required, and both groups must be on the same wavelength.

Fourthly, the capability of **process-oriented skills and training** refers to all initiatives that prepare managers and employees for doing their job and acquiring BPM knowledge and skills. Organizations need capable managers and employees, and should give them opportunities to follow BPM courses and constantly improve their BPM skills set. When they start thinking more deeply about business processes, they also reflect on process efficiency and effectiveness, and may be encouraged to think of creative solutions. This implies that also a digital skills set is required to form teams with an optimal combination of diverse skills for BPM-related projects.

In sum, these four BPM capabilities having the strongest relationship with digital innovation reflect that digital innovation has several objectives, e.g. making business processes more efficient and effective or creating an optimal work environment. The third and fourth BPM capability suggest that the corporate culture and human capital are driving forces or facilitators for digital innovation. Also [37] refers to the ultimate importance of people involvement, and [3] suggest that the human factor in business processes can make the difference. Since the first capability represents the organizational structure, evidence is given that BPM should rather be implemented top-down and so trickling down to the organization. Unfortunately, organizational aspects such as culture and structure are not included in many BPM maturity models [19].

6   Conclusion

Based on a survey with 131 West European managers, this study showed a positive relationship between BPM and digital innovation, as well as IT investments rather than IT outsourcing. The need for IT alignment, the decreasing costs for (new) IT, and the increasing role of the IT department may explain why organizations keep control over their IT infrastructure and (core) processes. Since this relationship turns out to be moderate, it emphasizes the complementarity between BPM and digital innovation as well as the risk that overprocessing and rigidity kills creativity. Hence, a balance needs to be found when organizations define their optimal level for BPM adoption. By focusing on BPM capabilities, four areas come to the foreground: (1) process-oriented governance bodies, (2) the Act-phase of the process lifecycle, (3) top management commitment and (4) process-oriented skills and training. They emphasize the relevance of human capital and organization-wide support. Since all BPM capabilities contribute to digital innovation to some extent, evidence is given of the usefullness of BPM maturity models if they consider the organization's culture and structure.

This paper is a first step towards a portfolio of pathways. A deeper understanding can be gained by looking at concepts beyond BPM (e.g. innovation processes or Lean Startups) or addressing the BPM capabilities in qualitative research. Further questions relate to: 1/the nature of the relationship (i.e. causality: "are organizations using BPM for better digital innovation, vice versa, or some other way?"), 2/the location of BPM groups (i.e. if BPM is initiated in and driven by the IT department, then the tested relationship may be tautological or self-referential), or 3/the actual problems (i.e., who is complaining and why: "are employees involved in innovation complaining about missing BPM expertise, or are BPM experts experiencing that their work is not needed?"). We thus open the discussion whether the role of BPM needs to change and whether the BPM-related capabilities need to be re-interpreted for the digital future.

Appendix A: The Profile of Organizations and Respondents

(See Tables 8, 9 and 10).

Table 8. The distribution of our sample for organization sector, using NACE codes (N = 131).

Sector	Frequency	Sector	Frequency
Agriculture, forestry, fishing	1	Financial, insurance	14
Mining, quarrying	1	Real estate	1
Manufacturing of products	37	Scientific, technical activities	6
Construction	9	Administrative/support service	3
Electricity, gas, air conditioning	2	Public, defense, social security	5
Wholesale, retail, vehicle repair	12	Human health, social work	5
Transportation, storage	8	Arts, entertainment, recreation	5
Accommodation, food service	8	Extraterritorial bodies	1
ICT	13		

Table 9. The distribution of our sample for organization size (N = 131).

Number of employees	Frequency	Number of employees	Frequency
1–10	25	501–1,000	10
11–50	22	1,001–5,000	16
51–250	25	>5,000	19
251–500	12	I do not know	2

Table 10. The distribution of our sample for respondent's seniority (N = 131).

Years	Total seniority	In current organization	With BPM involvement
0–5	12	54	87
>5–10	23	28	24
>10–20	23	27	14
>20–30	51	16	4
>30	22	6	2

References

1. Harmon, P.: (2016). http://www.bptrends.com/bpt/wp-content/uploads/2015-BPT-Survey-Report.pdf
2. Alotaibi, Y.: Business process modelling challenges and solutions: a literature review. J. Intel. Manuf. **27**(4), 701–723 (2016)
3. Van den Bergh, J., Thijs, S., Viaene, S.: Transforming Through Processes: Leading Voices on BPM, People and Technology. Springer, Berlin (2014)
4. Brynjolfsson, E., Hitt, L.M.: Beyond computation. J. Econ. Persp. **14**(4), 23–48 (2000)
5. Abrell, T., Pihlajamaa, M., Kanto, L., vom Brocke, J., Uebernickel, F.: The role of users and customers in digital innovation. Inf. Manag. **53**(3), 324–335 (2016)
6. Schumann, C., Tittman, C.: Digital business transformation in the context of knowledge management. In: ECKM Proceedings, pp. 671–675. Academic Conferences International Limited, UK (2015)
7. Fichman, R.G., Dos Santos, B.L., Zheng, Z.: Digital innovation as a fundamental and powerful concept in the information systems curriculum. MIS Quartely **38**(2), 329–343 (2014)
8. Schmiedel, T., vom Brocke, J.: Business process management. In: vom Brocke, J., Schmiedel, T. (eds.) BPM - Driving Innovation in a Digital World, pp. 3–15. Springer, Switzerland (2015)
9. Recker, J.: Evidence-based BPM. In: vom Brocke, J., Schmiedel, T. (eds.) BPM - Driving Innovation in a Digital World, pp. 123–143. Springer, Switzerland (2015)
10. Rosemann, M.: Proposals for future BPM research directions. In: Ouyang, C., Jung, J. (eds.) AP-BPM 2014. LNBIP, vol. 181, pp. 1–15. Springer, Cham (2014)
11. Dumas, M., La Rosa, M., Mendling, J., Reijers, H.A.: Fundamentals of BPM. Springer, Berlin (2013)
12. Hammer, M., Champy, J.: Reengineering the Corporation. HarperCollins Publishers, New York (2003)
13. Davenport, T.H.: Process Innovation. Harvard Business School, Boston (1993)
14. Dijkman, R., Lammers, S.V., de Jong, A.: Properties that influence BPM maturity and its effect on organizational performance. Inf. Syst. Front. **18**(4), 717–734 (2016)
15. McCormack, K., Johnson, W.C.: Business process orientation. St. Lucie Press, Florida (2001)
16. Hammer, M.: The process audit. Harvard Bus. Rev. **85**(4), 111–123 (2007)
17. Bronzo, M., Vilela de Resende, P.T., Valaderas de Oliveira, M.P., McCormack, K.P., de Sousa, P.R., Ferreira, R.L.: Improving performance aligning business analytics with process orientation. Int. J. Inf. Manag. **33**(2), 300–307 (2013)
18. de Bruin, T., Rosemann, M.: Using the Delphi study technique to identify BPM capability areas. In: ACIS Proceedings, vol. 42, pp. 642–653 (2007)

19. Van Looy, A., De Backer, M., Poels, G., Snoeck, M.: Choosing the right business process maturity model. Inform. Manag. **50**(7), 466–488 (2013)
20. Van Looy, A., De Backer, M., Poels, G.: A conceptual framework and classification of capability areas for business process maturity. Enterpr. Inf. Syst. **8**(2), 199–224 (2014)
21. Pöppelbuss, J., Plattfaut, R., Ortbach, K., Niehaves, B.: A dynamic capability-based framework for BPM. In: HICCS Proceedings, pp. 4287–4296 (2012)
22. Weitlander, D., Kohlbacher, M.: Process management practices. Serv. Ind. J. **35**(1–2), 44–61 (2015)
23. Trkman, P.: The critical success factors of BPM. Int. J. Inf. Manag. **30**, 125–134 (2010)
24. McCormack, K., Willems, J., Van den Bergh, J., Deschoolmeester, D., Willaert, P., Stemberger, M.I., Vlahovic, V.N.: A global investigation of key turning points in business process maturity. BPM J. **15**(5), 792–815 (2009)
25. Bucher, T., Winter, R.: Taxonomy of BPM Approaches. In: vom Brocke, J., Rosemann, M. (eds.) Handbook on BPM 2, pp. 93–114. Springer, Berlin (2010)
26. Nylén, D., Holmström, J.: Digital innovation strategy. Bus. Horizons **58**, 57–67 (2015)
27. Pilav-Velić, A., Marjanovic, O.: Integrating open innovation and business process innovation. Inf. Manag. **53**(3), 398–408 (2016)
28. Kemsley, S.: Emerging technologies in BPM. In: vom Brocke, J., Schmiedel, T. (eds.) BPM - Driving Innovation in a Digital World, pp. 51–58. Springer, Switzerland (2015)
29. Grenadier, S.R., Weiss, A.M.: Investment in technological innovations. J. Fin. Econ. **44**(3), 397–416 (1997)
30. Mithas, S., Tafti, A., Mitchell, W.: How a firm's competitive environment and digital strategic posture influence digital business strategy. MIS Q. **37**(2), 511–536 (2013)
31. Kane, G.C., Palmer, D., Philips, A.N., Kiron, D.: Is your business ready for a digital future? MIT Sloan Manag. Rev. **56**(4), 37–44 (2015)
32. Valdez-de-Leon, O.: A digital maturity model for telecommunications service providers. Technol. Innov. Manag. Rev. **6**(8), 19–32 (2016)
33. Nordstokke, D.W., Zumbo, B.D., Cairns, S.L., Saklofske, D.H.: The operating characteristics of the nonparametric Levene test for equal variances with assessment and evaluation data. Pract. Assess. Res. Eval. **1**(5), 1–8 (2011)
34. Box, G.E.: Non-normality and tests on variances. Biometrica **40**(3–4), 318–335 (1953)
35. Vickers, A.J.: Parametric versus non-parametric statistics in the analysis of randomized trials with non-normally distributed data. BMC Med. Res. Method. **5**(35), 1–12 (2005)
36. Shingala, M.C., Rajyaguru, A.: Comparison of post hoc tests for unequal variance. Int. J. New Technol. Sci. Eng. **2**(5), 22–33 (2015)
37. Hung, R.Y.-Y.: BPM as competitive advantage. Total Qual. Manag. Bus. Excel. **17**(1), 21–40 (2006)

Author Index

Alharbi, Amirah   88

Barbon Jr., Sylvio   55
Batoulis, Kimon   106
Bulpitt, Andy   88

Ceravolo, Paolo   55
Cho, Minsu   19
Christiansson, Marie-Therese   3

Damiani, Ernesto   55
del-Río-Ortega, Adela   19

Fernandez, Pablo   19

Garcia, Felix   127

Ivanchikj, Ana   36

Johnson, Owen   88

Kühnel, Stephan   71

Lübke, Daniel   36

Mancebo, Javier   127
Mandal, Sankalita   141
Meroni, Giovanni   160
Moraga, Maria Angeles   127
Müller, Carlos   19

Pautasso, Cesare   36
Pedreira, Oscar   127
Plebani, Pierluigi   160

Resinas, Manuel   19
Ruiz-Cortés, Antonio   19

Sackmann, Stefan   71
Seyffarth, Tobias   71
Song, Minseok   19

Torabi, Mohammadsadegh   55

Van Looy, Amy   3, 177

Weidlich, Matthias   141
Weske, Mathias   106, 141